The Radio Book™
The Complete Station Operations Manual

Volume Three:
Sales & Marketing

1995 Edition

Edited by: B. Eric Rhoads
Reed Bunzel
Anne Snook
Wendy McManaman
Vicky Bowles

Cover and Book Designed by:
Practical Graphics, Inc.
2090 Palm Beach Lakes Boulevard, Suite 903
West Palm Beach, Florida 33409

West Palm Beach, Florida

Streamline Press is a division of Streamline Publishing, Inc.
Copyright © 1995 by Streamline Publishing, Inc. All rights reserved.

No part of this publication may be reproduced, stored in a retrieval system or transmitted, in any form or by any means, electronic, mechanical, photocopying, recording, scanning or otherwise, without prior written permission of the publisher.

ISBN: 1-886-74502-1 (Vol. #3)
SET: 1-886-74503-X (3 Vol. Set)

Library of Congress Catalog Card Number:
The Radio Book Vol. #3 95-67241

Published by and distributed by:
Streamline Press
224 Datura Street, Suite 718
West Palm Beach, Florida 33401-9601
Phone 407-655-8778
Fax 407-655-6164

Printed in the United States of America

The Radio Book™
The Complete Station Operations Manual

TABLE of CONTENTS

Preface .... *B. Eric Rhoads* i

Chapter 1
Great Radio Takes Place In The Mind .... *Bill Burton* 1

Chapter 2
Analyze Your Station, Dominate Your Market .... *David J. Rogers* 5

Chapter 3
Truth Or Dare .... *Kathryn L. Maguire* 11

Chapter 4
Get $100,000 A Month .... *Chris Lytle* 15

Chapter 5
What Are Your Salespeople Worth? .... *Richard W. Chapin* 19

Chapter 6
Arbitron, Fishing And PMS .... *Ted Bolton* 23

Chapter 7
So What About Co-op, Anyway .... *Val Maki* 27

Chapter 8
Five Words Salespeople Fear Most .... *Alan Cimberg* 31

Chapter 9
Keep Your Customers .... *John Fellows* 37

Chapter 10
Cultivating Creative Genius .... *Dr. Philip J. LeNoble* 41

Chapter 11
Radio's Bum Rap .... *Walter Sabo* 45

Chapter 12
Typing People For Selling .... *David J. Rogers* 51

Chapter 13
A Whole New Ball Game .... *Chris Lytle* 55

Chapter 14
Divisions, Divisions .... *Val Maki* 59

TABLE of CONTENTS

Chapter 15
Those Dare Necessities.................................Ted Bolton..................63

Chapter 16
How To Be Your Station's Superseller....................John Fellows..................67

Chapter 17
Paying Attention To Retention.........................Chris Lytle..................71

Chapter 18
Say The Magic Words..................................Dwight Case..................75

Chapter 19
Trick Questions.......................................Dave Gifford..................77

Chapter 20
Temporary Solutions, Permanent Problems..............Gina Gallagher..................81

Chapter 21
Make Radio Tangible..................................Michael B. Hesser..................85

Chapter 22
Trial By Fire..Dr. Philip J. LeNoble..................87

Chapter 23
Is Your Station Burned Out?..........................Chris Lytle..................89

Chapter 24
A Niche In Time......................................Godfrey and Ashley Herweg..................93

Chapter 25
The Nurturing Edge...................................Mimi Donaldson..................97

Chapter 26
Turn Today's 'No' Into Tomorrow's 'Yes'..............John Fellows..................101

Chapter 27
To Kill A Vendor Department..........................Val Maki..................103

Chapter 28
Client Science..Dr. Sharon Crain..................107

Chapter 29
Dialing For Dollars....................................Pam Lontos..................111

Chapter 30
Getting Stores To Play Along..........................Kathryn L. Maguire..................115

Chapter 31
Becoming Self-ful...Mimi Donaldson...............119

Chapter 32
Radioís Little Instruction Book ....Chris Lytle 123

Chapter 33
Take Me Seriously Or Take Me Dancing..................Dr. Sharon Crain.............127

Chapter 34
Follow Up Or Foul Up?..Nancy Friedman..............131

Chapter 35
Use Software To Maximize Sales...........................John Fellows...................135

Chapter 36
Courtesy Sells..Jack M. Rattigan 139

Chapter 37
If Not CPP, What?...Bill Burton......................141

Chapter 38
Sales Warfare ....Jack Trout 143

Chapter 39
Leading A Turnaround..Gary Fisher......................147

Chapter 40
Donít Fire The Copywriter Yet! ....Judy Carlough.................153

Chapter 41
Know The Trades..Kathryn L. Maguire.........157

Chapter 42
Taking Control..Dr. Sharon Crain............159

Chapter 43
The Real Reasons Clients Buy..............................Chris Lytle 161

Chapter 44
The Furniture Boom..Dr. Philip J. LeNoble 163

Chapter 45
Grill Your Salespeople ....Chris Lytle 167

Chapter 46
Emotional Strategy..Ted Bolton......................171

TABLE *of* CONTENTS

Chapter 47
Understanding And Selling Lifestyles .... *Linda G. Brown* 175

Chapter 48
Copy That Sells .... *Pam Lontos* 179

Chapter 49
Furniture, Bridal And Bath .... *Dr. Philip J. LeNoble* 183

Chapter 50
What Radio Taught TV .... *Cliff Berkowitz* 185

Chapter 51
From Ink To Air .... *Pam Lontos* 187

Biographies 190

Index I

Back Issues .... ?

PREFACE

Radio is an ever-changing medium. As members of the Radio community, we must become students of the industry to grow our industry, our station revenues and our personal revenues. That, of course, is easier said than done. Time is a valuable and rare commodity. Nevertheless, without taking the time to invest in our growth, we risk becoming stagnant. Thus, we begin to regress gradually. To stay on the cutting edge, we must seek intellectual stimulation.

All too often we tend to believe that "there's nothing new in Radio." Perhaps old ideas do resurface with a new face. However, new ideas can change the way we operate — even if they're old ideas in a new package. This book is filled with ideas: Some are old; others are fresh and new. Even revisiting the basics is important. After all, your circumstances and experiences have changed since you learned the basics. Reviewing them can give you new ways of applying them or can refresh ideas you've long since forgotten.

We encourage you to use this book as a means of stimulating thought and building a better career, a better station and a better industry. This is not a one-sitting book, but rather a reference book. When you've got a problem, turn to the index and find a chapter relating to your need. You'll find it a valuable tool.

B. Eric Rhoads, Publisher

Chapter 1

Great Radio Takes Place In The Mind
Creativity Is The Key
By Bill Burton

Creativity is the magic word for success in Radio. A national advertiser (any advertiser) can spend a substantial sum of money on a Radio campaign — but if the creativity is lacking, the advertiser has lost a bundle and the campaign is doomed.

The absence of creativity is probably the major reason there is approximately a 50 percent annual turnover of national Radio accounts. Radio is not alone in the need for greater creativity. It is the make-or-break point on any national advertising campaign.

Radio's creativity is the advertiser's key to one of the great sales vehicles — the automobile — a Radio with four wheels. It can sell prospects right behind the steering wheel all the way to the point of purchase — moving products to the most captive audience in America.

A great Radio commercial demands greater talent and creativity than in any other medium. Why? Because you're only limited by the size of the mind and the imagination. For instance, close your eyes and visualize the following: the smell of a new car. Isn't that refreshing? You're probably ready to go out and start looking for a new car now.

The Tongue Can Paint What The Eye Can't See

Radio is an ideal sales vehicle to stretch the imagination as well as the mind. What better medium to sell the great aromas of perfume, shaving lotion, a warm vegetable soup for lunch or the smell of turkey and ham cooking? There is no way you could convert these wonderful aromas to picture or film — but the visualization in the mind can be overwhelming. All great Radio takes place in your mind. The characters and situations you identify with, the taste, smells, emotion, all come to life through the power of your imagination.

Bill Ludwig, executive vice president, creative director, Lintas:

Chapter 1
Great Radio Takes Place In The Mind

Campbell-Ewald, wrote a great article for our Detroit Radio Advertising Group titled: "See What I'm Saying." It does a masterful job explaining how creative can take advantage of the "Theater Of The Mind." Incidentally, Ludwig is the heart behind the very successful Chevrolet "Heartbeat" campaign.

Do It With Demos

In our push to get greater Radio creativity so we can be that much more successful in moving clients' products, let me suggest whenever possible that you produce the commercial into a demo tape. Recently, in putting on our "Get Me Radio" presentation to the advertising manager of a major national automotive account, the ad manager made the following comment at the end of my presentation. He said: "Geez, Bill, it's unfortunate, but when our agency presents Radio to our top management committee, many times it's a mock-up where the creative writers are voicing the spots and trying to get us to imagine how the commercial's going to come out. Whereas, if they're presenting television or print, they've invested a considerable amount of money in some excellent visual mock-ups."

Quite frankly, I was a bit amazed, particularly because so many Radio stations on the local level do a magnificent job of putting together spec spots that sell local advertisers. Let me suggest, henceforth, that all of you involved with national accounts or larger accounts do everything possible to encourage good Radio demo tapes. For the record, I went back immediately to the agency of the automotive client, met with the creative director and put forth a very strong pitch to do demos for their client. And he bought it.

Best Broadcasting Boutiques

Remember, top talent broadcasting boutiques are but a phone call away, and you can recommend them to agencies and clients. Chuck Blore, a legend in his own time — and mine — continues to do outstanding creativity, and may be reached at 213-462-0944. Jim Conlon and his partner, Bill West (with the "Radio Works"), are on the cutting edge, and may be reached at 713-789-3751. Dick Orkin, at 213-462-4966, is one of the best and also does a great brainstorming workshop for agencies, clients and stations.

Tony Schwartz, the master of political and thought-provoking creativity, may be reached at 212-581-5025. Interestingly, Schwartz also teaches a creative course on sound at the Harvard Business School from his studios on 56th and 10th Avenues in New York. He does it all by sound.

Let me suggest that all of you in Radio keep pushing for outstanding creativity in the utilization of our great Radio sales medium. Build a rapport with the creative talent at your agencies. Feed them good examples of good Radio. Promote the Mercury Awards. With the help of Jim Thompson, president of Group W Radio, and the RAB, it has the largest cash awards in the nation for great creativity. A $100,000 first prize, four $20,000 prizes and 20 $1,000 prizes. Help creative be creative.

Chapter 2

Analyze Your Station, Dominate Your Market
An Exercise In Strategy
By David J. Rogers

This chapter is designed to facilitate your thinking about some important elements of strategic planning. Specifically, it focuses on your views of your station's or group's: 1.) strengths, weaknesses, opportunities and threats (SWOTs); 2.) Core Competencies — your station's main abilities, the two or three things your station does better than anything else; 3.) key factors of success (KFS); 4.) ability to compete with rivals.

In the following exercise, you are going to identify your industry's and market's KFS (key factors of success) and SWOTs (strengths, weaknesses, opportunities and threats). You will also define and indicate how you'll put to full use your station's Core Competence. But before doing so, it is necessary for you to understand exactly what is meant by KFS, SWOTs and Core Competence.

- Identify your industry's and market's key factors of success (KFS), the two or three things that will make a Radio station in your market financially and competitively successful.

- Identify your station's or group's three main strengths, weaknesses, opportunities and threats (SWOTs).

- Identify your station's or group's Core Competencies, the two or three things you do better than anything else.

- If you correctly identify the KFS for Radio in your market and then make that KFS your Core Competence, you'll dominate Radio in your market.

SWOTS

The SWOT method was invented by professors at Harvard University and is often used in strategic planning. At the outset, identifying your station's or group's strengths, weaknesses, opportunities and threats may seem too easy, and for that reason you may try and race

Chapter 2
Analyze Your Station, Dominate Your Market

through it. But if you want to put it to best use, you must devote some real thought to the process.

Understanding your SWOTs is vital, because each and every one of your effective strategies will build on your strengths and take advantage of opportunities, while also overcoming weaknesses and threats.

To respond effectively to change in your environment, you must assess your groups or station's internal and external conditions now and in the future. Specifically, you have to know: 1.) your internal strengths and weaknesses; 2.) the external opportunities available and the threats the group or station faces.

Strengths and weaknesses are internal inside your operation. Threats and opportunities are external — outside your operation.

Possible Strengths: a special competence; you're a strong competitor; cost advantages; good staff; financial support; good reputation; market leader; good GM; excellent marketing abilities, sales staff, programming and other functions; value-added services superior to competitor's quality product.

Possible Weaknesses: lack clear strategic direction; under attack; poor equipment; poor management talent; bad at following through plans; weak image; poor marketers; financially weak; communication problems; not particularly skilled in any area; bad profit-making history; marketing research weak; little training and staff development; reactive, not proactive management.

Possible Opportunities: to serve new customer types; diversification; faster service delivery; develop new services; add-on complementary services; growth market; changing demographics; weak competitors; serve new advertisers.

Possible Threats: harmful government policies and regulations; emergence of new competitors; possible substitute products/services; slowing market growth; demographic changes; lifestyle changes; not ready for changes; vulnerable to recession; format popularity declining; advertisers shifting to other media; innovations by competitors.

Potential Functional Areas of the Group's or Station's Strengths and Weaknesses: Marketing — advertising, promotions, direct sales; Finance and Accounting; Personnel — productivity, efficiency, absenteeism, motivation, morale, job satisfaction; Management/Organization — planning/goal-setting, flexibility, interpersonal skills, control; Production; Product/Service Quality.

Example Of Station's Main SWOTs

S = We have an ability no competitor has.
W = A bad image.

O = New customers in the market.
T = A new competitor.

Core Competence

Core Competence is your group's or station's main ability; a capacity it possesses; the thing your group or station does particularly well. For example: customer service. Your station may have more than one.

Key Factors Of Success (KFS)

KFS are in the industry, the business. They are the two, three or four things in which a group or station must excel if it is to be financially and competitively successful. KFS refer to the things you must do particularly well.

There is almost nothing you can do that's more important than correctly identifying the industry's KFS because, invariably, without one exception in every industry in the United States, the competitor who identifies the industry's KFS and then makes it their Core Competence comes to dominate the industry. The KFS of brewing is distribution. Anheuser-Busch is the best distributor. Its Core Competence is the same as the industry's KFS, thus it rules the brewing business.

If you are right in identifying the KFS for Radio in your market and then make that KFS your Core Competence, you'll dominate Radio in your market. Remember, it's happened in every industry, without exception.

KFS vary from one industry to another and from one market to another. For example, a 7-Eleven's KFS are:
- Location — good store location.
- Products — the right range of variety and impulse products.
- Prices — higher than normal prices.
- Operating efficiency — poor management of a 7-Eleven will quickly erode the 3 percent pretax profit margin.

Start Exercising

Now that you know what's involved in identifying your SWOTs, KFS and Core Competencies, you may start the exercise. You will first list as many of your station's or group's SWOTs, KFS and Core Competencies as you can, then select the three main ones of each.

Remember, when it comes to selecting the three most important of each, the less you guess, the better off you'll be. Instead:
- Prove that what you've selected is true.
- Have other staff fill out the questionnaire and hammer out the truth.
- Go through this process at least two or three times. You may begin.

Chapter 2
Analyze Your Station, Dominate Your Market

Key Factors Of Success
What, in your judgment, are the Key Factors of Success in your industry and in your market?
1.) List as many possible KFS as you can on a separate sheet of paper.
2.) Select the three most important KFS for your industry and market.

EXTERNAL
1.) On another sheet of paper, list opportunities and threats to your group or station.
2.) Now list the three most significant opportunities and threats.

INTERNAL
1.) On another sheet of paper list your station's or group's strengths and weaknesses.
2.) Now list the three most important strengths and weaknesses.

Playing With SWOTs
You've identified the three most significant strengths of the station, weaknesses of the station, opportunities in the market and threats in the market. Ordinarily, this is where SWOT analyses end. But go further and mentally play with these scenarios:

Compare Your Internal ... With your External
- 3 Strengths + 3 Opportunities
- 3 Weaknesses + 3 Opportunities
- 3 Strengths + 3 Threats
- 3 Weaknesses + 3 Threats

• Strengths with opportunities is a very positive situation.
• Weaknesses with opportunities means you must turn those weaknesses to strengths to seize the opportunities.
• Strengths with threats — you can probably overcome the threat.
• Weaknesses with threats — you must buttress the weaknesses to offset the threats.

Core Competence
1.) List your group's or station's three Core Competencies — the three main strengths of your group or station, the three things you do particularly well.
2.) Write how, in your judgment, the group's or station's three Core Competencies can be used to improve your position in the marketplace. Be as specific as possible.

Are The KFS Your Core Competencies?
You've identified your station's Core Competencies and you've

identified the industry's and market's KFS.

For you to excel, they must be the same. There can be no difference between them.

For example, if you have concluded that community involvement is the KFS — and, in fact, you're right — community involvement must be your Core Competence. If it isn't, you have to develop it into your Core Competence.

Main Competitors' Strengths And Weaknesses

List what you know are your station's main competitors' strengths and weaknesses: competitor A, competitor B, competitor C, etc.

Analysis Summary Sheet

List your three most important: KFS, opportunities, threats, weaknesses, strengths and Core Competencies.

In analyzing the above, ask yourself:

1.) Are the KFS your Core Competencies? (They should be.)

2.) Of the opportunities you've listed: What prevents you from pursuing them? How could you take advantage of them?

3.) Of the threats: How can you eliminate them, or at least reduce them? If threats continue, what's the worst that can happen?

4.) Of the weaknesses you've listed: What do they keep you from doing? How could you get rid of them?

5.) Of the strengths, ask: How can we push hard to take full advantage of them? How can we develop them even more?

Chapter 3

Truth Or Dare
What You Think You Know Can Hurt You
By Kathryn L. Maguire

Falsehoods: All vendor dollars will come from the food business; vendor sales take too much time to pursue; you can only go after vendor business if you have a specialist on staff; it takes three to six months to get a vendor department going; a vendor director should do the footwork for the account executive; some station formats are better-suited to vendor sales than others; co-op dollars are not big enough to pursue; vendor means "over-and-above" co-op dollars.

Truism: You can close a manufacturer in 30 days.

A lot of "tall tales" about vendor sales exist in Radioland today. And there are tales that seem "tall," but are actually true. Here are a few.

Dare To Destroy The Myths And Go For The Gold

If you're going to go after vendor dollars, all of it is going to come from the food business. False. Admittedly, there is a lot of money in this category, but there are plenty of other lucrative categories to pursue. Automotive aftermarket, home improvement and clothing are good areas to look into.

You can close a manufacturer in 30 days. True. Ah ha! Did you think I would say false? The rule of thumb for a typical manufacturer program is three to six months, but programs have closed in less time. Just don't go in expecting manufacturers to close as fast as your traditional Radio accounts, or you will come off looking unprofessional to your prospect.

Vendor sales takes a lot of time, making it impractical for established account executives to pursue. False. First, think of manufacturers as additional prospects. If you are a very aggressive account executive, two new manufacturer appointments per week is excellent. An ideal weekly appointment level for adding vendor into your sales routine is one. And yes, there is more detail work involved — but don't forget that most of

Chapter 3
Truth Or Dare

this business comes in direct (more commission), at a higher spot rate (your boss will love you) and the average orders tend to be much higher than regular Radio orders.

You can't go after vendor business unless you have a specialist on staff. False. Although having a vendor director on-board is the ideal scenario, a lot of stations are adding substantial manufacturer-oriented new business dollars to their billing. Usually, the station has one or two account executives on staff who take it upon themselves to approach this business. And, occasionally, stations get special vendor sales training for their staffs to empower them to get those dollars.

It takes three to six months for a vendor department to get going. False. It takes three to six months for a vendor program to close, but you must give your department a year to develop. And year two is when you should really reap the benefits of your commitment.

A vendor director's role is to do the account executive's footwork. False. Think of your vendor director as a special local sales manager for a moment (although vendor directors are not typically considered management). This person works with your entire sales staff, just like the local sales manager does. This person trains and motivates the account executives, and helps them put the programs together. The account executives are accountable to the vendor director for weekly, monthly and yearly vendor goals, too. It is important that the account executives see this person as a specialist, but also as a teammate who's there to help them make money.

Some station formats are better-suited to vendor than others. False. Stations all over the country — with a lot of different formats, mind you — are making money in vendor sales. Take your pick, which one are you? Country, Talk, News, Sports, AC, CHR, Urban Contemporary, Jazz, Rock ...

Co-op dollars are not big enough to pursue. Vendor is where the money is. False. All those billions of dollars left unspent each year (that you have been hearing about most of your Radio life) often translate into millions of dollars in your very own market. One of the very best ways to get at the biggest co-op dollars is to work directly with the manufacturer sales rep for your area.

Vendor means "over-and-above" co-op dollars. False. Vendor is a synonym for manufacturer. If you asked a manufacturer where he kept his vendor funds, you would likely get a blank stare. There is really no such thing as vendor dollars. It's a name we attach to manufacturer money that is different from accrued co-op dollars. A little cheap advice? Try not to call this money anything at all. Just concentrate on creating a sales-driven campaign for your manufacturer prospects and let them decide from where the funding will come. (Hint: It could come from a key market fund, discretionary fund, case allowances, brand manager, etc.)

Did you get a few surprises? Hopefully, they are pleasant ones. Despite the former beliefs or misconceptions you may have had, just remember there is always money if the manufacturer believes they can make money with your idea.

Chapter 4

Get $100,000 A Month
Think New Dollars First, New Advertisers Second
By Chris Lytle

Pursuing new accounts is very tempting, especially as the cold reality of January and February sets in. But going after new accounts may not be the most efficient way to get new dollars in the station. Think new dollars first, new advertisers second.

Answer the following question, even if you have to get your bookkeeper to help you. (You are speaking to your bookkeeper, aren't you?)

$100,000 Question

"How many more dollars would your station have billed in November if every advertiser had run one more paid commercial than they did for every day they were on the air?"

To make this exciting — and easy to calculate — say you sent out 250 statements in November. Assume the average advertiser was on the air 10 of November's 30 days. Selling 250 advertisers 10 more commercials would give you another 2,500 commercials sold. At a $40 average media market rate, you would have billed another $100,000!

Easily, too. You don't have to write any more spec spots. You don't

---

Steps to take now for that extra $100,000 a month in billing:

- Determine number and frequency of billing advertisers on the air.
- Calculate money you would make if you sold every advertiser one more commercial per day.
- Set goals to increase average dollar per advertiser for each salesperson.
- Reward salespeople for upgrading current clients.
- Pay special attention to account attrition.
- Set goals in areas other than total billing.
- Require your salespeople to prepare pre-call objective sheets rather than call reports.
- Consider requiring minimum orders.

Chapter 4
Get $100,000 A Month

have to send out any more statements. You don't have to make any more sales calls. You don't even have to give salespeople extra commission for selling new accounts.

But don't take my word, or my hypothetical example for it. Do the math yourself. Present it to your salespeople in your next meeting. Then act.

Focus For The Year To Come

To get focused on big sales increases in the year to come, answer these two question: 1.) How much does a quarter-page ad in your local newspaper cost? 2.) How much money does the average advertiser on your station spend in a month?

You may be surprised to learn that a typical advertiser on your station spends about as much money in a month on Radio as they spend on the paper for one day. Why?

Because Radio salespeople are allowed to make small proposals to their clients for below-average dollars. Or perhaps because managers are sending so many mixed messages to their salespeople that the salespeople have lost focus. They no longer know what's really important.

What's really important is selling clients schedules that are big enough to work, and then running copy that consists of compelling offers and asks your listeners to take action.

Need an extra $100,000 a month in billing? Here are 10 action steps to take now:

1.) Determine the number of billing advertisers on the air and the average number of days each one is on the air.

2.) Calculate how much money you would have made if you had sold every advertiser one more commercial per day.

3.) Rant and rave about No. 1 and No. 2 for the next quarter.

4.) Calculate the average dollar per advertiser for each salesperson, and set goals to increase that average.

5.) Start celebrating and rewarding salespeople for upgrading current clients.

6.) Pay special attention to account attrition. Maybe you should concentrate on eliminating churn rather than focusing on getting new business.

7.) Set goals in areas other than total billing. Some examples: number of accounts on the air, average dollar per advertiser, number of new accounts, renewal accounts, seasonal accounts and number of long-term committed advertisers. (If you don't have our Account List Management System form, please call 800-255-9853 and the Advisory Board will send it out at once.)

8.) Monitor how many dollars your salespeople are asking for, rather than how many calls they are making.

9.) Help your salespeople prepare for every call they make by requiring written pre-call objective sheets rather than call reports. Pre-call planning is proactive. Call reports are reactive.

10.) Consider requiring a minimum order, or at least a minimum billable order. If it takes five calls to collect a little account, the sale is no longer profitable.

Focus on getting more productivity out of every call — and more billing out of every account that's already on the air. Think new dollars first, new advertisers second.

Chapter 5

What Are Your Salespeople Worth?

By Richard W. Chapin

Failure to believe in the product has ended many sales careers. When that happens among leaders, it is doubly hard for followers to get the job done. In Radio, most owners and managers design and run their stations presuming sales failure rather than expecting success. Radio has an outstanding listenership. Even the smallest AM daytime station has more than adequate audience to satisfy the financial ambition of any merchant with a worthy product or service.

When ads fail, people have rejected the advertiser's message, not the station's. Even a "wrong" demographic can be sold on the old bank-shot concept — use one group to motivate another. "Kids, tell mom to ..." is one of the first ads I can remember.

Our product is advertising, not audience. That simple oversight by many owners and managers has cost millions of dollars. When sales dip, we tend to try to fix the audience instead of the sales department. We try a new format, fancier equipment, a taller tower and computers of all kinds to raise the billings.

- Failure to believe in the product has ended many sales careers.
- When sales take a dip, we tend to try to fix the audience instead of the sales department.
- When we quit adding, training and upgrading salespeople, we announce that growth is finished.
- Commission-based pay is associated with chancy products and implies permission to not sell once commissions have satisfied financial goals.
- Our entry-level people start at about $1,000 monthly and receive two weeks' training — tops. We are sending financial privates to sell generals.
- When we believe enough, we will build advertising companies, not just Radio stations.

Chapter 5
What Are Your Salespeople Worth?

We forget that people make sales calls, "things" don't.

When we quit adding salespeople, we announce that growth is finished. Our justification is usually some discussion of "points of diminishing return" or "spending too high a percentage of income on sales." In reality, owners and managers are tired of searching for, training, monitoring and firing salespeople, because then they have to start over again. Three out of four entry-level sellers never finish the first year. To get six means hiring 24. Setting selling standards means you will have turnover. Strong salespeople require strong structure.

The fact is that those who get through the first year will level off to a minimum monthly billing, and then you are only at a minimum rate. The "stars" who go higher are lost to other industries or by promotion into management. In this twilight zone they sell and, in the leftover time, manage others who sell. Few humans have two great talents, and sometimes both jobs suffer.

The easy way out is to adjust the rules of production and behavior to avoid turnover. This is one way our leaders show their lack of faith in selling and advertising.

With low rates, cut rates and soft collections, you have proof of low product belief at the top. If something has true value and is respected by you, nobody gets it cheap or is allowed to pay slow. Between emphasizing friendship-based selling and fearing the loss of an account through firm and fair collection methods, we can't make the case that we believe in our product.

That "a thing is worth what is paid for it" is an old business law. Offers and appraisals do not count. Cash-in-fist sets the true value to both buyer and seller. Commission-based pay is associated with chancy products. Most successful industries pay a substantial base and then bonus for added production. This allows them to set firm selling performance requirements for the base wage. Selling rules (calls per day, demo tapes for day, annuals, etc.) must be followed. Commission-based pay implies permission to not sell once commissions have satisfied financial goals. This explains why successful sellers "level out." The amount paid for sales is one thing, but the method of payment may be equally important. Worse, it signals that leadership doubts the product's value.

The average Radio advertising salesperson bills about $9,000 monthly, which provides an average of $1,350 monthly at 15 percent commission. Our entry-level people start at about $1,000 monthly and receive two weeks of training — tops. Most successful decision-makers called upon by our people make vastly more than that. We are sending financial privates to sell generals. The shortest military boot camp is 13 weeks. A recent study of sales compensation shows that the average trainee pay is about $2,000 monthly and involves 14 weeks of training. The average salesperson in the business world earns $3,800 monthly. Senior sellers make far more than that. We are low on wages and provide one-seventh the training. Turnover may be the good news. We may suffer more from turnoff. Those good and successful sales candidates who never apply go where the bucks are. The training is thorough and the rules and standards are strong enough for their selling pride. When was the last time a real top seller switched from AT&T or IBM to Z100? When we believe enough, we will build advertising companies, not just Radio stations.

Chapter 6

Arbitron, Fishing And PMS ...

By Ted Bolton

Recently Arbitron announced the imminent introduction of a revolutionary advancement in audience measurement techniques. The technology is called "Pocket People Meter," not to be confused with the Popiel Pocket Fisherman. Furthermore, Arbitron has made it clear that although the technology is called PMS (passive measurement system), it is in no way designed to simulate the discomforts and irritability caused by the better-known syndrome.

Even though the technology is easily two years away from market introduction, this announcement has raised a host of important questions from broadcasters across the United States.

Wired For Sound, Ready To Proliferate

As a listener, you will really never know whether or not you will be a Pocket People Person, because Arbitron will designate PPPers randomly. However, as a broadcaster, you can be certain that ratings will be part of your daily life. You see, the Pocket People Meter has

- The Passive People Meter will deliver overnights so you can feel the impact of this new measurement system every single day ... that is, if you want to.
- Pocket People Persons will carry a beeper-like device that will "hear" inaudible acoustic sounds transmitted by a Radio station within "earshot." Earshot has not yet been defined.
- Since "earshot" will replace "unaided recall," it may be better to put all your money into roving remote vans that can broadcast your station all over town.
- Realize that the PPM could become a fashion statement.
- Lobby against future sales and development of personalized stereo Walkman-type devices. Earshot does not apply to them.

continued

Chapter 6
Arbitron, Fishing And PMS

the capability of delivering overnights so that you can feel the impact of this new measurement system every single day of the week, all year long.

If you have been selected by Arbitron to be a PPP, then you will be asked to carry a beeper-like device that can actually "hear" the inaudible acoustic sounds transmitted by a Radio station within "earshot" of the PPP. Earshot has not yet been defined to the best of my knowledge. The People Meter will then store all this earshot data and send it back to Arbitron via modem or when the meter is returned.

The Really Important Questions

Q: Since the term "earshot" will replace the term "unaided recall," should we consider putting all of our money into roving remote vans that broadcast our station?

Yes, as long as your blasting station is within earshot of the Passive People Meters. Start thinking now about how to "disarm" competitive Radio vans that will also be blasting their stations across town. Day-to-day engineers may be replaced with G. Gordon Liddy types who truly understand the ins-and-outs of electronic sabotage, signal jamming and competitive wiretapping.

Q: Will wearing a Pocket People Meter become an important fashion statement?

As we all know, the presence of a beeper on the belt can connote a certain sense of fashion, importance and urgency for the person wearing it. *The Wall Street Journal* recently reported that beeper sales have skyrocketed among teenagers who are clamoring to wear them as fashion symbols.

Arbitron is also considering a smaller People Meter that could be worn as a wristwatch or small lapel pin. Arbitron should definitely initiate discussions with Swatch, ESPRIT and Polo to stay in tune with the latest fashion trends. Think of the embarrassment and humiliation you'd feel wearing an unfashionable People Meter.

Q: Should we lobby against the future sale and development of personalized stereo Walkman-like devices?

Yes. You see, "earshot" does not apply to Walkman-like headphones. Unfortunately, the beeper-like People Meter is unable to pick up this earshot signal. However, Arbitron is considering giving all PPPs special headsets that will be able to register listenership and avoid the obvious audio paradox.

Q: Are there any new positioning strategies and promotional tactics Radio managers should consider once People Meters are introduced into their market?

The latest inside intelligence is going with some dramatic on-air positioners designed to influence PMS. They include: • The one station that just HAS to be loud. • Back-to-back LOUD hits! • The louder the better. • Continuous loudness ... all day, all night, all the time. • We kill people who wear Walkmans.

Q: Can this kind of rating service hurt my dog?

Potentially. Reportedly, there has been some concern that the People Meter emits an ultrahigh frequency heard only by dogs. Fortunately, the People Meter only accepts frequencies and does not generate any kind of signal that will make your dog go crazy. But anything's possible.

> *continued*
> - **New positioning strategies could include such gems as "Back-to-back LOUD hits!" or "Continuous loudness ... all day, all night, all the time."**
> - **PPMs could potentially harm your dog.**
> - **Pack your bags and move to your dream small market ... PPMs will only be used in larger markets.**

How To Avoid What's On The Horizon

Whether you like it or not, little can be done to slow progress. There may be one solution, however. Arbitron has stated that the People Meter will only be used in larger markets, while smaller markets will continue with the diary. Solution?

Pack your bags and move to your dream small market or, if you're already there, go to bed tonight feeling secure that you will never have to deal with PMS, G. Gordon Liddy or a neurotic dog again.

Chapter 7

So What About Co-op, Anyway?

By Val Maki

Co-op. Co-op. Co-op. There, I said it! I don't care anymore if anyone thinks co-op is a bad word. The truth is, co-op advertising has been relegated to the ranks of the unsophisticated sales type because generally a co-op program is comprised of an established set of rules and guidelines, thereby erasing most of the mystery of the craft.

> - When approaching retailers, manufacturers and wholesalers, ask a lot of pre-planned questions.
> - Understand fraudulent billing.
> - Get the proof of performance right the first time.

Consider this ... whenever someone is trying to sell something to someone else, and someone or something can help that someone sell that something to someone else, there is an opportunity — if you can help that someone sell that thing. That's the basis of what you do in any vendor, co-op or sales increasing program, regardless of the level of distribution or class of trade. To that end, co-op is no different from any other type of vendor program. And there's nothing sophisticated about it, although sometimes there is something "crafty."

The Focus

If an individual or staff at a Radio station chose to focus only on co-op programs, learning them backward and forward, they could satisfy their new business goals, as well as account upgrade goals, every year.

One of the first benefits from pursuing co-op dollars only is that doors would open to so many kinds of funds anyway, because you would be talking to the sellers and buyers — the right contact people.

Call Strategy

Here are a few proven tips for approaching manufacturers and retailers:

Chapter 7
So What About Co-op, Anyway?

Retailers. Sell them on an idea. Then suggest a couple of ways to fund that idea — maybe from a co-op program, a manufacturer's rep with whom you've been working or someone you know would be amenable to listening to a sales-increasing idea.

Ask them for help, direction and ideas on a manufacturer who may be more promotionally oriented and more likely to provide support. It really does work to do research on their products' accruals and distribution chains. Undoubtedly, some possibilities for funding a program will turn up. There are companies that help with this research, too!

Manufacturers. Ask for direction. Tell them you have something in common. You call on the same accounts, therefore you have something to talk about. Use examples of retailers with whom you've worked, and programs you've completed, for credibility. Use their language and ask a lot of questions.

Distributors/Rep Firms/Wholesalers. Every case is different, depending on their participation. Ask a lot of questions.

Know The Right Stuff

You can't turn around in this industry without tripping over some retail terminology. Ingest some. And co-op and general marketing terms as well.

Have a list of basic questions handy for the phone calls and appointments. Also, talk to someone who has made co-op sales and ask them what the main objections are.

Even though you know some "stuff," it's best to be humble:

"If I'm not mistaken, you seem to be doing pretty well with your product in Phar-Mor ..."

"Pardon me if I sound one step removed here, I don't know your business as well as you do, but ..."

"Let me ask you something ..."

"I need some help ..." or "I need some direction ..."

But temper this with bouts of credibility. For example: "Let me ask you something. We need to find a product to feature exclusively in an annual promotion we do with Dayton's Department Store. Can you give some direction?" Remember, you need to know some stuff, not all the stuff.

Legalities

You absolutely need to know about fraudulent billing, still a hot subject and often requested, innocently, of course. You may be surprised at the depth of what constitutes fraudulent billing. You may even want to carry a copy of the dissertation with you when you make calls, in case there is a gray area. Here again, use humility in your posturing. "You know, you wouldn't think that would be the least bit illegal, but ..." (Some retailers still request promotions that are lotteries. Although it's basic, it wouldn't hurt to review it with your staff.)

Proof Of Performance

You must provide the correct proof of performance for co-op schedules the first time, as soon as humanly possible, after a schedule. Anything else is unacceptable if you want to sell and resell co-op advertising.

A major advertiser recently told me that they may not do Radio again because the paperwork was "out of hand." It seems a huge percentage of the Radio stations they worked with did not provide the simple proofs of performance that correctly matched the retailer's various products advertised. Their disappointment was not with Radio as a means of promotion, but as an unreliable partner in follow-up service.

If you can educate yourself and the staff involved in becoming experts in proof of performance and aiding clients with their co-op claims, you will get the business every time. The only complaining you'll hear from the client will be about the other stations who are no longer getting their business because of incorrect handling of paperwork. So pick up a co-op information source today and make just one call. Doors will open.

Chapter 8

Five Words Salespeople Fear Most

By Alan Cimberg

We all know there are seminars that improve the skills of salespeople and teach them how to deal specifically with buyers. But did you know there are seminars, attended by buyers, that help them handle salespeople? They really do exist — I've been to them.

While eavesdropping on one of these seminars, I overheard some buyers talking about you — the salespeople who call on them. What I heard spoke volumes about the weakness of salespeople through the ages. One buyer said: "I could take the most confident salesperson in the world and reduce that person to a stuttering, blubbering idiot with five simple words."

You know what those words are: "Your price is too high." You've heard it. You've felt its impact. It stops most of you in your path just when you thought the sale was going smoothly.

Why is this objection constantly thrown at salespeople? Why do buyers use it so often? The first and most obvious reason is it may be true. Your price may be too high. More often than not,

- The primary reason buyers claim your price is too high is they are looking for a better deal.
- If you hear "Your price is too high," it means you have failed to justify your price. Justify your price by creating value.
- In giving presentations, it is essential that you give your buyers all the benefits that are relevant to their needs. Do not cut your presentations short.
- Make sure you take the time to care. And show your customer that you do.
- Find out what the objections to your price are — then you can deal with them.
- Negotiating is tough — quote only your best price and be flexible only if your customer is flexible.

Chapter 8
Five Words Salespeople Fear Most

however, your price is not too high. After all, stations stay in business only by remaining competitive.

The primary reason buyers claim your price is too high is they are looking for a better deal. That's their job. They have to test to see if there is a better price to be had, and they often get it when you cave in.

When you hear the words "Your price is too high," you are being told that you have failed in one essential aspect of the sales process; that is, you have failed to justify your price. You do that by creating value.

What is taking place in the minds of the buyers is a weighing process. They are evaluating what you have told them about the features and benefits vs. the price you have quoted. If the scale tilts too far in the price direction, they object with "Your price is too high." If the scale tilts toward value, they buy. If the scale is balanced, they ask for a better price.

The Benefits Of Benefit

Quite often, salespeople fail to build up the weight of the perceived value of their product. Hence, when the weighing process takes place, they lose. So, it is absolutely essential to give your customers all the benefits that are relevant to their needs. This requires extensive product knowledge and the discipline not to trim the presentation through the repetition of it. Too often, salespeople get tired telling the same old story, so they cut more and more pieces out of their presentation. In effect, they shortchange their prospects themselves. We must remember that the prospect is hearing it for the first time. So make it "first-time fresh" — that's what the stage performer does even for the 900th performance.

You have to realize what you are doing when you leave out details of the presentation. Imagine I want to sell you an audiotape of one of my speeches. The price is $500. You would be appalled and tell me: "Five hundred dollars! Not only have you lost the sale, you've also lost your mind!" And you would be right, because I left out one essential part of my presentation — the part in which I tell you that each audiotape comes with a diamond-studded, gold Rolex watch. If I had told you that, your weighing process would have been different and my price would no longer have been too high. The sale would have been saved by that "minor" detail — one that added value to tip the scales.

Take The Time To Care

Another reason salespeople fail to justify their prices is they don't care about their customers. They don't take the time to uncover the needs of their prospects. As an example, let me tell you what happened to me the last time I shopped for a new car.

I went to a local auto showroom and was greeted by a man who didn't even bother to get up from his desk. He said: "May I help you, sir?" Now there's a great opener for an automobile salesman. I felt like saying: "Yeah, I came in to buy a tie. Have you got something in brown to match my eyes?"

When I revealed the startling fact that I had come in to buy a car, he took me over to a sedan and lifted the hood. He pointed proudly to the engine and said: "Look." I looked, but didn't see what he wanted me to see. I know nothing about engines and care even less. Then he took me to the trunk. He opened it and pointed to the roomy compartment. "That will hold five pieces of luggage," he said. Well, that disqualified me immediately. I only own two pieces of luggage, and wasn't about to go out and buy three more pieces just because the trunk could hold them.

Let's look at what was happening in that interaction. Did the car salesman give a hoot about me? Of course not. He was interested in only one thing — making the sale. He didn't care about adding value to the sale or assuring me that the dealership would still be in business after I drove off the lot.

How do you think I felt about the salesman? Horrible! And in a business that revolves around the strength of personal relationships, a salesperson cannot afford to alienate a prospect by being so impersonal.

If salespeople do not find out what my needs are, they cannot customize their presentation to those needs. When I say to them: "Your price is too high," they will be unprepared to answer my objections. Instead of answering with relevant comments, they will have to generalize about the features and benefits of the car. It just won't work!

What Does It All Mean?

What does it mean when someone says: "Your price is too high?" There is only one way to find out — ask! Use whatever words are comfortable for you, but tactfully find out the reason for the objection. Then, and only then, can you address the issue intelligently.

There are several reasons for being told a price is too high. You might hear someone say: "Well, you're asking $5,000, and I only have $4,500 in my budget." That's one of the easiest objections to handle. Notice that the price is not too high, but the customer just doesn't have the money. Your job now is to work with your prospects to find a way to make the sale possible — help them find a creative way to pay for it, sell them a less expensive package or show them that their budget is unrealistic for what they need and maybe they'll find more money.

When asked to elaborate on his objection, another customer might say: "Your price is too high because the competition charges $100 less."

Chapter 8
Five Words Salespeople Fear Most

This happens most often when the salesperson is coerced into giving his price before presenting a complete picture of the relevant benefits. When this happens, it is too easy for the prospect to dismiss your product with an unfair comparison to your competition.

We have all met prospects who were so busy they just wanted to know the bottom-line price. It's easy to cave in and give them what they want, but don't do it. It would be unfair to both of you. A price is meaningless without a value attached to it, and that is the salesperson's job — to convey value. So be polite, but insist on telling your prospects what they are getting for their money.

Why would a comparison based solely on price be unfair? Most often, the comparison is between apples and oranges. So be sure to point out the differences to your prospects. These differences should be described in terms of features that will be benefits to them vis-a-vis their specific needs.

It is important to have done your homework on your competition before a comparison is made. If you have, you will be able to answer their objections quickly and realistically. Part of a realistic comparison is the discussion of the bottom-line costs that buyers understand so well. Although a competitor's price may be lower, in the long run your products may cost less. For example, you may pay the transportation costs or extend better term payments.

Remember this: When comparing prices, you have to justify only the differences in price, not your total price. So if you charge $500 and the competition charges $450, you should only address the $50 difference. If all else fails, you can reduce it to the ridiculous. "Fifty dollars, when amortized over the life of this contract, is only $2.50 per spot ..."

Negotiating

Negotiating is a tricky business. It is a custom in some countries to haggle over price — a process I find uncomfortable. For example, imagine someone offering you a watch that looks pretty good. You ask: "How much?" They say: "$100." You offer them $10. They say: "I'll take it."

After the sale, how would you feel? Some people might feel exhilaration because they talked the person down 90 percent. Someone else may feel bad because they're wondering how much lower they might have gotten the price if they had tried. In either case, it would make me worry about the quality of the watch. Is it a $100 watch that I got for $10, or a $5 watch for which I paid too much?

Negotiations are risky, so always tell your prospects that you have quoted your best price and you can only be flexible if they can be flexible. Never cut your price unless you get some kind of concession from the buyer. For example, your customers may agree to increase the size of their order or accept a less targeted schedule. You can always remind your customers that "you get what you pay for."

Be A Pitcher, Not A Catcher

Here's some advice on handling objections in general. Don't answer an objection and then pause, waiting for the next one to be hurled at you. Instead, move on with your presentation or, better yet, ask for the order.

Above everything else, if you do not feel good about the price, nothing will help. I suggest you visit three of your good customers and ask them: "Why do you buy from me and my station?"

Chapter 9

Keep Your Customers
Grow Sales
By John Fellows

Quick! What's the easiest sale to make? Salespeople everywhere will say: "A sale to an existing satisfied customer, of course."

The Quality Imperative

We don't keep enough customers. It's no secret that keeping customers is the key to growth, that satisfying your customers is the key to keeping them and that servicing them above the norm (more importantly, above their expectations) is the key to satisfying them. Yet, when you examine the customer attrition rates of Radio stations across the country, you see a clear and disturbing picture of an industry unable to keep customers.

> - Service is the key to satisfying customers.
> - Satisfying customers is the key to keeping them.
> - Keeping customers is the key to growth.
> - We spend too much time and effort getting customers and not enough time servicing, satisfying and keeping them.
> - Use these 20 tips and you'll keep more customers.

Plenty of reasons for the problem exist, and many of them need just one basic, time-tested, street-level solution: superior customer service. Hey, don't take my word for it. Management gurus throughout business have latched onto this same notion, as if revealing some great new truth, profoundly labeling it "the quality imperative"... so it must be true.

The Wrong Focus

What's our focus? "Yeah, yeah, yeah," you say, "we tried customer service once and it didn't work." Don't laugh. Some stations haven't even tried it once. In Radio, we tend to focus on how to get customers and neglect how to keep them. Think about it: When was the last time

Chapter 9
Keep Your Customers

a rep came back from sales training and said: "Wow, that was great! I learned 20 ways to keep my customers!"

Enough editorializing. Here's a salesperson's starter list for superior customer service. (Suggestion: Get together with your entire staff — salespeople, personalities, administrative staff — and brainstorm ways to provide superior customer service to your customer groups, advertisers and listeners. Then implement the ideas.) As you go through the following list, remember that every time you say to yourself: "Who's he kidding? I don't have time for that!" you're reducing your value to your customers, and your ability to keep them.

Twenty Ways To Keep Your Customers

1.) Stay in touch with your clients, even when they're not on the air: off-season, between scheduled flights, etc.

2.) Contact your clients when they are on the air. Call them with the scheduled ad times and estimated listenership at those times (use hour-by-hour numbers if available). At the very least, do this the first day of a flight.

3.) Send your prospects and clients articles and materials in which they may have an interest ... about their business, related business, etc.

4.) Ask open-ended questions (who, what, when, where, why, how ...) and listen carefully. Their problems and ideas can be your opportunities for superior service.

5.) Always give your customers a cassette and typed copy of their ad.

6.) Send them thank you notes ... when they pay a bill, for a meeting, order, etc.

7.) Do what you say you're going to do, and when you're going to do it. Then do it!

8.) Treat your customers as individuals, not as a client category. Don't prejudge or stereotype.

9.) Be prepared to objectively advise your customers how to buy other media (if you're competent to do so). Point out other media's strengths if and before you point out their weaknesses.

10.) Explain your station's systems — order flow, traffic, billing, production, etc.

11.) Give your customers written proposals ... and get signatures or broadcast insertions for all orders.

12.) Give your customers sales leads for their business. Hand out their business cards to your contacts — it only takes one sale to one of your referrals to make your customer really happy with you.

13.) Give your customers more than the norm. Exceed their expectations whenever you can!

ARE WE STUPID, OR WHAT?

I went into McDonald's recently, gave my order and handed the girl a $50 bill. "Shit," she said, and she went into the back room to find some change.

Yesterday, dressed in jeans and a T-shirt, I went to Marshall Fields to buy five dress shirts. I unwrapped a shirt and the saleswoman, not liking the way I looked, shouted to me from behind the cashier: "Sir! Sir! You cannot try on shirts unless you have been measured. Bring me that shirt!" I brought her the shirt, then left.

Carson Pirie Scott has shirts just as nice.

At my bank, there are two tellers on payday. Consequently, I have to wait in line for 20 minutes to cash a check. And if I ask them to check my balance, print a money order or perform some other kind of service, they charge me. In trying to make it easy on themselves, they forget the purpose of their jobs.

A plethora of statistics — and common sense — tells us that a customer who is treated badly will tell at least 10 people of his or her experience.

Does anyone understand the importance of these statistics?

Yes. The cab driver who jumps from his cab to open the door when he sees me struggling with three suitcases; he'll get a bigger tip. The delivery girl who delivers my papers, placing them neatly upon the sidewalk in front of my door; she'll get a bigger tip. The car salesman who remembers my birthday and writes me a card. He'll get my return business.

Why do so many companies promise so much when they smell your money, but forget you once they've cashed your check?

Once again, look at the facts: A recent survey finds that within 10 years, 81 percent of the average company's customers drift away. Only 13 percent leave because they are dissatisfied with the product or service. But 68 percent of customers leave because employees are indifferent or rude. And most firms spend five times as much to get a customer as they do to keep one.

What's the problem here: Are we stupid, or what?

— *Brian Ragan, president of The Bottom Line Communicator*

14.) Take care of mistakes quickly. Nothing works more insidiously against you than a neglected customer problem.

15.) Get involved with them socially when possible, in business and community groups, etc.

16.) Meet with their accounts payable person. Walk them through the billing from your company. Acquaint them with your accounts receivable manager.

Chapter 9
Keep Your Customers

17.) Give your customers a list of key people at your station. Managers, administrative, receptionist, etc. Emphasize support staff functions.

18.) Demonstrate your accountability. Keep them posted on progress of projects.

19.) Ask them: "In your opinion, what could we do better?" and "What should we do that we're not doing?"

20.) Ask them for back issues of their industry magazines. Learn about the current key issues in their industry, region, company.

Service, Service, Service

Everyday, all around the country, you and I work hard to locate, qualify, develop and present to advertising prospects — all this effort in the hope of turning them into customers. Too often, we invest so much time in those prospecting activities we don't leave enough time for the service activities that keep those customers. It's no secret that service is the key to satisfying our customers, and satisfying customers is the key to growth. So, as the accompanying article says: "Are we stupid, or what?" Let's do it.

Chapter 10

Cultivating Creative Genius

By Dr. Philip J. LeNoble

If you haven't looked lately, the Radio industry is changing rapidly. LMAs are springing up everywhere in attempts to shrink the pools of red ink. Agency buyers have turned sales associates into commodity brokers who compete for the lowest rate ... and get it. And national business is a joke.

What will be Radio's competitive advantage for the rest of the century? Billy Grooms, general sales manager at WMFX, Columbia, South Carolina, believes the missing link is creativity.

The Creative Edge

"The agency business is going down," Grooms says. "If you're not creative, you'll go down."

With all the competition out there, stations selling similar formats and everyone with a TapScan rating of No. 1 in the 25-54 market, who does the local-direct buyer buy?

- Creativity is the driving force behind Radio and could be the missing link to making Radio competitive when sales are down.
- Try helping to write and produce a commercial. It could do wonders for the client and make the station look great.
- Move away from selling Tapscan, Strata and other ratings-related presentations. Train the right hemisphere of the brain and think creatively.
- Creative salespeople don't look for prospects, they create them.

One answer might be, whoever delivers the commercial with the biggest bang. Why not help programming by helping to write and produce a commercial? It could do wonders for the client and make the station look great.

After all, creativity is the driving force behind Radio. And advertisers care less about who is No. 1 than who can help produce the most

Chapter 10
Cultivating Creative Genius

effective commercial to make their business stand out.

Besides my recommendations that a salesperson have drive, courage, intellectual and cognitive skills, the most rewarding attribute is creative potential. We know how to sell Radio, but do we know how it works? Do we teach the client that creativity is the key to advertising's effectiveness? Or do we just know how to slam dunk another two-week order to make this month's projections?

Training The Imagination

Perhaps, instead of sending our salespeople to another Holiday Inn seminar, we should enroll our sales associates at our local colleges and universities in a class on creative writing or copywriting. Although some sales managers only want their associates selling, and not writing copy, I believe the creative position can be a competitive advantage. You go in with an idea, and you win. Go in with the lowest rate and a TapScan run, and you're dead!

To earn the highest income in the industry, you must train the right hemisphere of the brain and think creatively. The right brain is the nucleus for imagination, invention and visualization. With the seed of imagination, you can fertilize the mind of your prospect. You can help the prospective client see and project actual customers visiting their store, fulfilling the needs and wants you have created in the commercial message. With the forces of imagination, you can be miles ahead of the competition; without it, miles behind.

A professional salesperson who tries to be imaginative is also curious. Curious sales professionals consistently strive to find out the how, when, where, who and why in getting the prospect to listen. How am I going to get them to listen to my presentation? What will it take to generate and hold their interest? Without the elements of curiosity, we cannot open new doors to the mind. The more you use your imagination, the more interested your prospect will be. Involvement makes it easier for them to see how you are helping them solve their problems. Your words build color into the presentation and develop ideas for the actual commercial message.

Creativity Sells

Move away from selling TapScan, Strata and other ratings-related presentations. Let your mind be the palette and your tongue the brush. Listen to the words you use to make the presentation, the words you write for a commercial. You can be rich in short words that tease the taste, make glad the vision, tickle the nose and please the ear. There are words you can hear like the swish of silk, the rattle of a taxi, the rustle of the bushes, the crashing of the surf. With words, you can make a presentation or create a commercial that sounds and feels like music on a scale.

I have always followed a basic path: Adapt ideas that others have found successful; form an informal think tank or brainstorming session with others on the sales team. Finding and using new ideas to create sales boosts earning power.

I discovered a list of useful methods for generating ideas in Jones and Healy's *Miracle Sales Guide*:

1.) Spend 15 minutes a day looking for ideas.
2.) Read more. Non-readers start far behind the line in getting ideas.
3.) Talk as often as you can with people who are better informed on subjects than you are.
4.) Write more letters to clients. Say something new in each letter.
5.) Learn by talking to other salespeople.
6.) Travel more often: You see new things, you get forced out of your non-thinking rut. (Get out on the streets!)
7.) Cultivate older salespeople and listen to them. Just because there's snow on the roof doesn't mean there's no fire in the furnace.
8.) Experiment with your own ideas. How can you tell they will work if you don't experiment with them?
9.) Adapt ideas from others. Go to the most pat and profitable ideas others have proved and use them.
10.) Consult with the experts. Whenever you have a chance to talk to the very best in the field, grab it. They'll always be eager to share their knowledge.
11.) Test your assumptions.
12.) Challenge the rules.
13.) Ask "what if" questions.
14.) Play with the problem.

According to Roger Von Oech, author of *A Whack On The Side Of The Head*, there is no one answer.

Creative salespeople don't look for prospects, they create them. In poor economies, when retailers are frightened, the big earners create survivors. Who wins the race isn't as important as who finishes. They become the leaders.

Chapter 11

Radio's Bum Rap
Time for Change
By Walter Sabo

You're smart. What's wrong with these statistics: Ninety-four percent of households own more than one Radio (according to the RAB); 68 percent of a working woman's media day is spent using Radio (RAB); 25 percent of all listening is in-car (according to Veronis/Suhler Media Report); 29.1 percent of all listening is at work (Veronis/Suhler); 13 percent of all U.S. workers are men with wives who do not work (U.S. Dept. of Labor). In addition, more than 54 percent of listening occurs outside the home, making Radio the most effective mass medium for reaching working families. Yet the NAB reports that 59 percent of stations lost money in 1991. And only 10 percent of all advertising dollars are invested in Radio, according to the RAB.

As the population moves into cars and offices, so does Radio. In-auto listening has risen 3.1 percent, compounded annually since 1985. At-work listening has risen 2.6 percent, compounded annually during the same period. Radio pervades every auto, store, office, factory and home in America. For almost 70 years, no

- The biggest advertiser complaint about Radio is lack of follow-up. Hire at least one sales service executive to handle a variety of client services.

- Get honest about commissions: Take the cap off commission earnings or raise salaries and establish a subjective bonus for performance.

- Raise your rates. Radio delivers precise audiences. Therefore, Radio should charge a premium for that service.

- Get rid of the Macho Mentality. More than half of all AEs are women, and almost half of all sales managers are women.

- Take 10 percent of the budget used to promote the station to listeners and invest it in marketing to current and potential clients.

continued

Chapter 11
Radio's Bum Rap

medium has matched the needs of the public as effectively as Radio.

Come Out Of The Dark Ages

It is time to recognize that the economic "problem" with Radio is not ratings or programming or the number of stations, it is the management of sales. While programming efforts are progressive, sales and marketing methodologies are shockingly similar to those used 30 years ago. Our programming strategies against other media are remarkably effective; our sales and marketing efforts often are not. Two examples:

First, local cable revenue for 1991 was $721 million, according to the Cable Ad Bureau. What rating service is used to sell local cable programming? None. The only numbers they sell are household penetration of cable and household profiles.

Second, despite their problems, most local newspapers gross more than all Radio stations in their market combined. The classified section of your daily newspaper generates more annual revenue than your Radio station.

It's time to bring the sales and marketing department to the level of the programming department. Here are a few simple steps toward that goal:

1.) A reason that the medium garners only about 10 percent of all advertising revenue is that Radio account executives spend only about 10 percent of their time selling. This is not the AE's fault. Their time is spent in endless sales meetings, lining up appointments, copying copy, mailing dubs, faxing schedules, collecting debts ...

The biggest advertiser complaint about Radio is lack of follow-up. Superior sales service is required if we are to compete with other media and maintain current clients. Hire at least one — two would be more effective — well-paid sales service executives (SSE). Not a glorified secretary. A person trained in handling complex service issues on the phone. (Airline telephone ticketing agents would be a perfect source of applicants.)

The SSE would have the authority to handle all follow-up with clients: Mail dubs. Send out copy. Adjust schedules. Plus, the SSE would handle the personal details: Send cards and gifts to clients on appropriate occasions. Take care of the ticket and goodies requests.

Pay the SSE a salary plus commission on all repeat business. As Federal Express and Japanese businesses can attest: Service, Service, Service is the way to build business.

Make the business office 100 percent responsible for collections. Since the business manager must approve credit, the business manager — not the AE — should have a salary tied to collections.

2.) Get honest about commissions. Most stations' commission systems are actually discretionary bonus plans. Once a top biller earns "too much," their commission is cut or their good accounts are taken away. Shifting accounts demoralizes the salesperson and erodes advertiser confidence in the medium. Either take the cap off commission earnings or raise salaries and establish a subjective bonus for performance.

> *continued*
> - Sales myths are killing the Radio business. Forget them.
> - If it ain't broke, break it. Review the entire sales structure and open up the possibility of doing everything in a new way.

Niche For Sale

3.) Capitalism works on two basic principles: The more expensive it is, the better it is; and that the price asked is basically fair. A Cadillac is perceived to be more desirable than a Chevy because it costs more — even though you know that they are essentially the same car.

Advertisers pay top dollar for niche marketing vehicles. Radio invented niche marketing. Radio delivers precise audiences. Therefore, Radio should charge a premium for that service. *Gourmet* magazine charges more than $30,000 for a full page and delivers less circulation than most top 50 market stations' cume. They sell their niche.

Radio talks niche but continues to sell at bargain basement prices and volume discounts. When a retailer paying $2,000 a week in newspaper ads is offered "much more audience" for half the price on Radio, he thinks: "What's wrong with Radio? If it's better, why is it cheaper?" In other words: Raise the rates. Sell your niche. Sell results. Being the low-cost media leader is not working.

Eliminate the grid card. It is human nature to look at a grid and want the lowest price. Major retailers have told me that the grid card confuses them. And that confusion makes them think they are being had.

Pay commission on rates. The most effective way to raise and maintain rates is to base the commission structure on rate rather than volume. In the '80s, this was practiced at the RKO Radio stations with spectacular results.

4.) Get rid of the Macho Mentality. More than half of all AEs are women, and almost half of all sales managers are women, yet a macho mentality prevails: "I can sell anything." "I don't need any help." "The sales meeting is every morning at 8."

To recruit bright, winning AEs, dump the 8 a.m. sales meeting mentality. There is nothing heroic or necessary about going to work with farmers. Employees with young families will be loyal to the company

Chapter 11
Radio's Bum Rap

that offers flex time. Old Charlie may respond to the toughen-up 8 a.m. sales meeting, but Young Barbara may not.

Programming has evolved effectively because most program directors are open to help, suggestions and new ideas. Apply those attitudes to finding new revenue.

Using objective, external research, learn why your clients buy what they buy. Markets with 30 to 40 stations plus 60 to 100 cable options demand a fresh review of advertiser needs and perceptions. For a truly sobering experience, sit in on a pitch made by a magazine or major TV station to an agency or important client. You will see slides, audio effects, animated charts and four-color leave-behinds. All advertising is intangible. Other media do much more to appear tangible.

The macho mentality of "I can sell anything without help" results in Radio salespeople going on the street with dated leave-behinds and tacky presentation covers, and that's about it.

Invest In Client Confidence

5.) Take 10 percent of the budget used to promote the station to listeners and invest it in marketing to current and potential clients. Every dollar invested in reaching potential clients before your AE meets with them will enhance the likelihood of closing a sale. Every dollar invested in keeping a current client confident in your station saves money in turnover and reselling. How to invest the money?

• There are trade publications for every industry, such as *American Banker* or *Beverage News*. No Radio station has ever advertised in those publications. Imagine the impact if yours did. Buy an ad, then buy a subscription to those trade publications. Then your AEs will have a better knowledge of their clients' businesses.

• Bring outstanding guest speakers to your market to discuss non-Radio management and demographic trends. Invite all advertisers. Make your station the focal point for vital information.

• Establish a weekly media newsletter just for your market.

• Send clients excellent premium items for their offices. Let's get out of the key chain, coffee mug and poster business.

• Bombard potential clients with direct mail and presell telemarketing. Every time clients see your station's name, they become more familiar with it. And that familiarity inspires the confidence to buy.

• Invest in sales training seminars — not only for your sales team but also client sales teams. Then, you become partners.

• Show up at trade shows and conventions of major advertisers. Your station's presence will be noted.

Old School, Old Rules

6.) **Myths that kill.** It is possible that sales progress has stalled because most general managers come from sales.

When a PD tries to convince the sales-oriented GM to accept a new idea, the GM says: "Prove it. Get me research." If the PD can prove the value of a new idea, it usually gets done. Many GMs have minimal programming experience and, therefore, few programming myths to impose upon the PD.

Not so with the sales manager. When the sales manager goes to the GM with a new sales idea or a better way to organize the sales department, the GM has years of past sales experience to draw upon. The key word is "past."

The GM who comes from sales sometimes comes with a fondness for all of the sales myths, such as: our billing is down because the market is down; all time buyers are 24 and don't know anything; time buyers don't understand Radio; we shouldn't raise rates too fast; we'd raise our rates except that (name any competitor) doesn't; the rep is screwing us; the RAB is only for small-market stations; the RAB is only for major-market stations; P&G will never buy Radio.

Torch the myths. The way it's always been done does not work.

The major market GMs who have a hard-core history of solid sales performance, highest rates, no spiffs and low turnover in accounts and personnel never worked one second in a sales department; all understand the power and value of the product because they come from programming.

7.) **If it ain't broke, break it.** Review the entire sales structure. It makes little sense that 10,000 Radio stations organize their sales personnel essentially the same way: a structure established more than 40 years ago.

Meet with your sales manager. Open up to the possibilities of doing everything in a new way. Bring in experts from other sales/service industries. Brainstorm. No new idea is a bad idea.

Chapter 12

Typing People For Selling
Breaking The Mold
By David J. Rogers

Like every other field of human endeavor, sales is cluttered with false beliefs. Fortunately, there is enough research available today to shatter those stagnating misconceptions. For example:

You've got to make a positive impression in the first two minutes. Research repeatedly shows that prospects will like you best of all if they dislike you at first but change their minds later on.

Hire only sales types. Newer research reveals that the salesperson's personality is meaningless. The similarity between the salesperson and the prospect is what creates sales. A high-powered "sales type" might never sell to an introverted, close-mouthed prospect.

We need more sales training. Probably not if it's the traditional kind. Untrained individuals who possess one particular quality can outsell extensively trained senior salespeople who lack the quality.

Scores of research-validated insights support each of these five real keys to selling success: Communication, optimism, match, counseling and motivation.

- The five keys to selling success are: communication, optimism, match, counseling and motivation.

- Highly developed communicators who sell for a living can come quite close to developing relationships with 90 percent or 95 percent of their prospects.

- Optimists sell 37 percent more than pessimists.

- Account assignments are significant. The more alike the salesperson and the prospect are, the greater the likelihood that the prospect will buy.

- Typologies are lazy selling. They limit the sales rep's understanding of the whole client.

- Without motivation, the potential greatness of the salesperson will not be achieved.

Chapter 12
Typing People For Selling

The Five Keys

Communication. Sales representatives who are trained on certain principles of communication will easily outproduce reps trained in any sales program available anywhere. Highly developed communicators deal successfully in any interpersonal situation 95 percent of the time. Highly developed communicators who sell for a living can develop relationships with 90 percent or 95 percent of their prospects.

Optimism. More than 100 studies involving 15,000 people have shown that optimistic people are happier, healthier and more successful than pessimists.

They are also excellent salespeople. Research shows that optimists sell 37 percent more than pessimists. In one sales study, a company experimented with 100 optimists who had failed standard tests of salesmanship and originally were not hired. They immediately sold 10 percent more than the company's average. Positive thinkers who had never received any sales training outsold senior mid-level account executives who had attended numerous sales courses.

Optimists succeed interpersonally because they expect to succeed and they enjoy the job. In one study of prospects who were successfully sold, 96 percent of the prospects responded: "The salesperson enjoyed his job." Ninety-eight percent said: "He enjoyed talking to me." The startling fact was that only 20 percent of the buyers thought the salesperson represented the best company.

Perfect Pairs

Match. Research shows that account assignments are by far the most significant key to successful selling.

The more alike the salesperson and the prospect are, the greater the likelihood that the prospect will buy — and the higher the closing rate. Age, height, income, political opinions, religion, smoking habits, ethnicity, status, attitudes, interests, values and personality all affect how closely the seller and buyer can relate.

In light of these results, it is useless to seek one all-purpose "sales type." The ideal type of salesperson depends specifically on the client.

A more intensive matching relates to behavior. If the prospect talks very slowly, the best sales rep talks very slowly, etc.

Counseling. A somewhat popular approach to Radio sales training involves teaching reps how to fit their prospects into a small number of types or styles. All the account executive has to do is identify the prospect's type, then adjust the presentation to that type.

Typologies are lazy selling, and they ignore each prospect's rich and varied individuality. They also tend to focus on one overriding characteristic and fail to recognize that a person can change considerably from one minute to the next. They limit the sales rep's understanding of the whole client.

The skills of the masterful salesperson are closely akin to those of the skilled psychological counselor. If there is one thing the counselor tries never to do, it is to type the person. The counselor attends to the differences that make the client unique.

The salesperson trained on "types" takes the "type" for reality; the prospect becomes an abstraction. The salesperson who avoids typing the person but focuses on the prospect can produce a high number of sales.

Motivation. A person might acquire highly developed communication skills, become optimistic, match up with prospects and customers, and learn to use the skills of the professional counselor. But without a powerful urge to action, to contact and to meet, the potential greatness of the salesperson will not be achieved.

Can you train people to be motivated? Traditional thinking might suggest that you can't. But research suggests that the answer is a resounding yes.

Chapter 13

A Whole New Ball Game
By Chris Lytle

The Colorado Rockies and the Florida Marlins recently chose their players from the rosters of established major-league baseball teams, assuring that they started their first seasons with major-league players and prospects. Neither team planned to win a World Series for many years, but they will field competitive teams.

The draft worked this way: Each established team could protect 15 players from the draft. The rest of their big-league and minor-league players were left unprotected.

Use the principles of the expansion draft to shake up your account lists and create a flurry of new sales activity.

Better Prospects For Rookies

Too many new salespeople are calling on "The Charles Darwin Account List," accounts that veterans have determined have no potential. Meanwhile, the veterans hoard decent prospects, whether they are calling on them or not.

While the fit will survive and ultimately thrive despite the "Charles Darwin Account List," your station's sales suffer because the best prospects aren't being seen regularly.

The Marlins and the Rockies will at least start out with big-league

- Major-league baseball's expansion draft is the perfect model for building account lists.

- Let your salespeople protect their current billing accounts and 10 prospects that show promise. Let each salesperson, starting with your newest hire, "draft" the rest.

- Increase your station's billing by cutting the number of accounts each salesperson sells and services.

- Due to Radio's longer selling cycle, making multiple calls against fewer accounts puts sales on the books.

Chapter 13
A Whole New Ball Game

prospects, if not superstars. Unlike many Radio rookies, they have a chance to win.

Add By Subtraction

Anybody on your team who is calling on more than 50 advertising decision-makers has a hunting license — not an account list. The reality is that 80 percent of the business on the typical account list comes from eight to 14 accounts. Further research confirms that most salespeople are making their quotas with anywhere from 16 to 24 advertisers.

Here's how to shake loose some accounts:

• Announce an account draft. Let every salesperson protect every client they have had on the air and collected money from during the past 90 days. You may be generous and allow each salesperson to protect another 10 or 15 accounts that show promise. That's it.

• Every unprotected account is posted on flip charts. Let every salesperson on your team draft accounts. Let your newest hire have the first pick, your second-newest the second pick and so on, with your veterans choosing last. Continue until every salesperson has 50 accounts. Stop the draft.

• You now have a huge list of unassigned accounts with potential. Create another list of 50 accounts and hire someone to call on it. You may even want to hire two or three more salespeople. They will be easier to motivate, because they won't be calling on accounts that "have had a bad experience with your station" and "have all the business we can handle."

Concentrated Effort, Higher Yield

Optimum effective scheduling gets better results because it helps advertisers achieve a frequency of three against a significant percentage of your station's audience.

Let's assign 200 accounts to Jones and 50 accounts to Smith, and require each of them to make 10 calls a day. At the end of the month, Jones and Smith have each made 200 sales calls. Jones has made one call each on 200 accounts. Smith has made four calls to each of 50 accounts. By "concentrating the force" of sales calls against the target group, you can bet that Smith outsold Jones. Jones has a bigger list, but Smith has more sales on the books.

Farmers have used this principle for years. It's more profitable to increase the yield from the same acreage than to buy more fields. Salespeople who limit their calls and focus on 50 or fewer accounts find ways to increase dollars per advertiser.

Selling more to relatively fewer is more efficient and more profitable for you and ultimately the advertisers.

The Added Bonus

The expansion draft is a misnomer in that it actually cuts lists down to manageable size. That causes increased concentration on fewer accounts. Sales management's job is to deploy salespeople to sell and service the best accounts and prospects.

Like the Marlins and Rockies, salespeople have a chance to be competitive. They are calling on clients they want to be calling on.

The worst opening line in Radio sales is: "Your account has just been assigned to me." Use an "expansion draft," and even new people can start a meeting with a new client like this: "You were my No. 1 draft pick. It's good to finally meet you."

Chapter 14

Divisions, Divisions
By Val Maki

Individual account management is difficult to measure in vendor sales because we sometimes lack objective means. It is always interesting to hear someone who has just taken on a manufacturer account, perhaps a multi-divisional account like Procter & Gamble, announce their ownership of the account as though it were a singular pursuit. In truth, calling on a Procter & Gamble could mean "tens" of decision-makers and influencers.

The levels of sales management, marketing and overall brand management for any one manufacturer account requires a road map. Furthermore, the norm in any sale involves multilevel decisions. The same is true in doing vendor business with a retailer. Three basic steps — profiling, infiltrating and maximizing — can help you make strides in managing your accounts.

And remember, you can't possibly do a good job in account management with oodles of vendor accounts, so don't try. Choose your accounts wisely, and use an account profile to keep them in better focus.

- Multilevel decision-making is the norm in most types of sales ... and you need to cover all of the levels.
- Use an account profile form for every manufacturer and retail account.
- Approach and infiltrate the account according to the levels on the profile form.
- Maximize your exposure with an account by using the profile system. And choose your accounts wisely.

Profiling: Chart Your Strategy

The best road map for managing your accounts begins with an account profile. Use it to clarify all levels of influencing and decision-making in a manufacturer's or retailer's organization, and as the primary

Chapter 14
Divisions, Divisions

way to develop strategy in account penetration. Use any linear/diagram-type forms you want for the profile (or call me and I'll send you mine).

For The Manufacturer: Include any local, regional, divisional/zone and national levels of product sales for your market, such as manufacturer's rep, district sales manager, regional sales manager, etc., in the profile. Also include the national sales manager, brand managers and marketing managers for each division.

Allow space for any notes on local influencers at the trade level and in different classes of trade (perhaps covered by different sales representatives). This is important in situations when you are getting nowhere with a sales rep in one class of trade on a product that you think is a good match for your Radio station. Should you give up on that account for now or try the sales rep who calls on convenience stores? Or should you call on the district sales manager? It's best to start at the highest level of sales in your local market. With an organized account profile, you always have another option on whom to contact.

For The Retailer: Use a profile that will keep track of everyone from the CEO to store managers. If you use a department store, for example, account for all departments, the merchandising and buying staffs, the in-house advertising and PR departments, the advertising agency and any leased departments. You might also note key products' manufacturer reps on the profile, as they too can be influencers on your programs.

Infiltrating: Follow Your Chart

Infiltrating is simply using the account profile forms to prioritize and approach accounts. You can choose your contacts and with whom to set appointments. You can choose who will just receive mailers, letters, press releases, etc. ... for now.

A Manufacturer Example: The district manager for Helene Curtis acts as the key accounts sales rep in your market. You've discovered that Helene Curtis has new facial products. You call the district manager and find out that she is motivated to discuss one of your sales-generating programs. Had she not been, with her help or by calling Helene Curtis' headquarters in Chicago and finding out who the regional sales manager is, you could have filled in a few more spaces on your account profile form ... simply more levels to infiltrate.

A Retailer Example: You profile the major department store in town and find many levels to infiltrate. You decide to pick five departments that could be good matches for your station. For example, cosmetics. But you also keep track on your profiling form of the levels of merchandise management — merchandise manager, divisional merchandise manager, etc. In profiling a department store recently, 25 possible influencers or decision-makers were noted.

Once you have infiltrated an account through an appointment or an actual completed campaign, you can maximize that contact, all the while using your account profile as a guide.

Maximizing: Cover And Choose

Refer to the profile from an account folder at any time during the sale and ask yourself: "Is there someone else in the hierarchy who should be contacted at this point? Do I need to cover any other influencing bases? If the program has been completed, can I send cassette tapes of the ad to key personnel? Press releases? Recaps? Pictures? Have I held a recap meeting with the appropriate contacts?"

These are all ways to maximize your hard-won account penetration. And, along with the account profile, they will pave the way for future programs and great account relationships.

Chapter 15

Those Dare Necessities
By Ted Bolton

Football fans watched with anticipation in 1992 as the Dallas Cowboys won their way toward the best turnaround in recent NFL history. It came as no surprise that the basic elements of good management, risk-taking and long-term vision produced a winning football team. These same elements also produce winning Radio stations.

When Jerry Jones bought the down-and-out Cowboys in 1989, the team had a dismal 1-15 record. Yet Jones was optimistic. "You take something with this kind of history, even though it is down at the time, and get it productive, get it competitive, and you can see the results," he told *The New York Times*. He knew he was buying a franchise with inherent history and emotion. He built the team from there.

> - Vision, commitment and financial infusion can make the difference between a winning station and a loser.
> - Invest in talent. Hiring and firing are key management decisions that can turn your station around.
> - Plan on spending as much, if not more, than your competition.
> - Take risks. The risky solution could be the right solution.

Many Radio stations have the same kind of history and emotion. They might be only temporarily wounded. They might only need the kind of vision, commitment and financial infusion that Jones provided for the Cowboys.

Put Your Money Where Your Talent Is

Two key hires led the Cowboys to the Super Bowl. The first was coach Jimmy Johnson, who had led the Miami Hurricanes to a national championship and gained a reputation as a calculated risk-taker. It was

Chapter 15
Those Dare Necessities

---

the kind of risk-taking that Jones admired when he offered Johnson the job, amidst one of the most unpopular management moves, firing Tom Landry. But he banked on Johnson and made a statement about his total confidence in his new leader.

Johnson then went on to make the second key hire. "Stability and continuity in ownership and coaching are two of three key ingredients to success in pro football," Jones said. "The third is a superior quarterback." Troy Aikman was drafted in 1989 and became one of the NFL's most promising quarterbacks.

I know of not one successful Radio station that does not have in place its own version of the Johnson/Aikman combination. Without the best manager and PD, the chances of making the Radio Super Bowl are slim to none.

Risk Makes Right

The right thing to do two months into your first season as new head coach is trade away your only Pro Bowl player, Herschel Walker, right? Well, the answer proved to be absolutely right. In return for Walker, the Cowboys got five players from the Minnesota Vikings and six choices in future drafts.

So often in the risk-avoidance, safe playlist world of Radio, risk-taking has become a forgotten necessity of winning. Risk-taking sends a message to your employees that management is in this for the long-term win. As Jones points out: "In any organization, you can't sell short the need that things are going to get better."

Invest To Win

There is a cost associated with winning. Calculate the total out-of-pocket expenses for the leading station in your market. Then determine how badly you want to be No. 1. Because, to be No. 1, plan on at least matching, if not exceeding, their expense line. You cannot compete in Radio by spending less and expecting to win.

Jones, after only a few months into ownership, fired many of his middle managers. He then poured the savings back into players' and coaches' contracts for long-term security. Then he put more money into sales and marketing programs that lured more advertisers into the Cowboy camp. By securing talent and focusing on marketing, the Cowboys watched ticket sales soar and luxury boxes sell out altogether.

Jones borrowed heavily to acquire lease rights and management control for Texas Stadium. He moved his wife from Little Rock to Dallas. "I didn't buy the Cowboys to make money; I have money." He put everything on the line to prove that he was in the game and intended to stay there.

His example should come as a lesson to jittery owners terrified by a single downturn in an Arbitrend. Stick to the plan and work it through to completion. In Radio, as in football, you can't execute a plan halfway or halfheartedly. Believe in your franchise, your management team and the day-to-day talent that bring your property to life. Take intelligent risks and play for the long haul. If you are a believer, your team will turn into a believer and a winner. After all ... isn't that why you play?

Chapter 16

How To Be Your Station's Superseller

By John Fellows

> "I didn't accept their offer because it was a lousy list." Ever hear a sales rep use that excuse for turning down a position at a station? Ever use it yourself? Or how about this one: "Man, it looked like a great list when I interviewed, but it turned out to be dog meat."

- **Top-billing lists are built, not inherited.**
- **Do what the sales masters do.**
- **Apply the "Three Ps": Be pleasant, positive, persistent, and know when to move on.**
- **Be methodical, thorough.**
- **Stay focused on your goals.**

One of the biggest delusions is believing that luck is involved, as in the familiar: "Boy, is Joe/Jane ever lucky! They got a great list — all the best accounts — and I'm stuck with the slim pickings." Zig Ziglar calls that kind of attitude a "loser's limp." The fact is, almost no superseller's list is anything extraordinary when they first get it. It's not luck, but hard work that makes a great list.

Grow Your Own

In local Radio sales, there is only one way to get a top-billing list. Build it yourself. Oh, sure, you can inherit a list with a track record of performance; many people have. But inheriting and maintaining or, better yet, growing a top-billing local list takes lots of time, energy, skill and persistence.

Your success depends only on your ability to follow in the footsteps of others. While it can be tough to fill the shoes that made those footsteps, it takes only a dose of unwavering determination to become a "lucky" superseller.

An Arabic proverb suggests: "Follow the tracks of the fortunate person and you will come to fortune." All you need to do to succeed in

Chapter 16
How To Be Your Station's Superseller

sales, or in life, is use the success tips you read, listen to or watch. Do what the sales masters do. You don't have to adopt the specific selling styles of Tom Hopkins, Zig Ziglar, Brian Tracy or Tony Robbins, but you should be open to their ideas, try them out and adapt them to your particular situations.

Mind Your Ps

Apply the "Three Ps" of selling: Be pleasant, positive, persistent. You must remind yourself to be pleasantly persistent, positively persistent and persistently persistent. No one ever gets anywhere with sarcasm, anger, resentment or any of the other negative emotions.

Keep this in mind: 8 percent of all sales are made after the fifth call. Most salespeople give up on or before the third call. Remember also that your prospects aren't just waiting anxiously for you to come walking through their door. You must first prove to them that you are willing to work for their business before they'll let your product work for their business.

An important part of being persistent, though, is to know when to call off the dogs. You can't win over everyone you want. Don't waste time or energy pursuing prospects that your gut tells you aren't going to buy from you. Don't let your ego get in the way of the best interests of your station or your prospects. Turn the prospect over to someone else and move on.

Be methodical and thorough. Have a plan and stick with it. At the same time, avoid the tendency to "put on the blinders." Be ready to re-evaluate and modify the specifics of your plan to accommodate new information and situations while remaining true to the concept.

Goals And Inspiration

Make your goals attainable, challenging, chronological, measurable and divisible (into bite-sized pieces). Once you've set your goals, stay focused on them. As you near one goal, set the next one to avoid the deflating "what now?" syndrome.

Keep reminding yourself what your goal is, how you will reach it, when you'll reach it, who can help you (nobody ever attained their goals without the help of others), where your goal will take you and, most important, why you're striving for your goal.

Surround yourself with inspirational people, books, tapes and posters. Perhaps you'll want to do as I have and post this Brian Tracy quotation on your wall: "Do you have the discipline, the control, the perseverance, the persistence and the determination to hold your goal, your ambition, your aspiration clearly in your mind long enough for it to come to reality?"

If you ask yourself that question often, and answer it confidently, truthfully, in the affirmative, you will become your station's superseller. Regardless of the list you have today.

Chapter 17

Paying Attention To Retention

By Chris Lytle

If you're a good sales manager, you know in a heartbeat how much you billed last month. If you're a great sales manager, you will know four other vital pieces of information:

1.) Which clients were on two months ago who didn't advertise last month?

2.) What percentage of your station's billing is "churning" — or failing to repeat — each month?

3.) Which salespeople have the best client retention percentages?

4.) How do they do it?

Knowing how much you are billing is one thing. Now it's time to focus on how you are billing it. In this uncertain economy, sales managers are working to control the amount of churn on station account lists for one simple reason: Repeat business is more profitable than new business.

- Great sales managers ask "How are we billing?" instead of "How much are we billing?"
- The profits in Radio come from repeat customers, because they are more efficient to serve.
- It is easy to calculate churn as a percentage of last month's dollars and last month's advertisers.
- Salespeople who know you're paying attention to client retention may become more sensitive to this issue themselves.

Familiarity Breeds Efficiency

Customer service expert Frederich Reichenheld observes that the more times customers come back, the more efficient they are to serve. In Radio, this is true for many reasons:

1.) Once the salesperson has gathered the basic data, each meeting is shorter and generally more productive.

Chapter 17
Paying Attention To Retention

2.) Once the copywriter has developed a basic positioning statement and knows the client's preferences, the commercials take less time to write.

3.) Once an announcer has produced several commercials for a client and flagged a specific piece of music, the announcer's productivity increases.

4.) Once your bookkeeper has a working relationship with the

Churn Calculation Worksheet

STEP 1:
My accounts on the air last month:

Name of Company	Amount Billed
AABKO Mortgage	$800
American TV	2,357
Arnold's Soda Shop	445
Betsy's Box of Gifts	520
Chi-Chi's	1,200
Don the Muffler Man	660
Fortune Magazines	410
Hardy Hardware	777
Joe's Pizza	480
Linkon Auto Parts	1,600
Menson's Auto	910
New Horizons Health	1,466
Newstand Daily	600
Schwegler Lanes	2,250
Sunny Savings & Loan	1,222
What's New Books	560
Total sales last month	**$16,257**

STEP 2:
My accounts on the air this month:

Name of Company	Amount Billed
American TV *	$1,550
Arnold's Soda Shop *	610
Best Buy Wholesale	1,187
Chi-Chi's *	550
Civic Center Playhouse	650
Coliseum Tickets	2,280
Fantasy Gifts	218
The Tire Changer	456
J.J. Maxx	1,200
Joe's Pizza *	250
Linkon Auto Parts *	965
Madison Magazine	440
Momma's Deli	1,206
Olds World Auto	4,800
Our Own Hardware	882
Park Town Civic Group	610
Total sales this month	**$17,854**

STEP 3: Identify repeat business. (Indicated by *.)

STEP 4: Add up the dollars billed THIS month to businesses that are carryovers from LAST month. This is your REPEAT BUSINESS. **$3,925**

STEP 5a:

Total of last month's billing	$16,257
Less repeat business	—3,925
= TOTAL DOLLAR CHURN	$12,332

STEP 5b:

% of billing churned
.758
$16,257) $12,332
Last month's $ sales   Total $ churned

STEP 6a:

No. of accounts on air last month	16
Less No. of repeat business accounts	— 5
= TOTAL NO. OF ACCOUNTS CHURNED	11

STEP 6b:

% of accounts churned
.687
16) 11
No. of accounts   Total accounts
on air last month   churned
*Should be less than 20%

client's bookkeeper, the collection process goes more smoothly.

5.) With a core group of advertisers, it's easier for you to make projections.

6.) Customers who are getting a solid return on their advertising investment tend to be less price-sensitive than brand new prospects who don't know what they are getting into and have "heard stories" about the flexibility of Radio rate cards.

Calculating Churn

Look at the sample of the Churn Calculation Worksheet. On the surface, you see a sales increase of $1,597 from one month to the next. We know how much the salesperson billed — but look at how they did it. Only five advertisers repeated from one month to the next. Twelve advertisers are no longer on the air. They have lost 70 percent of the advertisers and 76 percent of the billing from the previous month.

Are you still happy about the $1,597 increase in sales?

With more attention to client satisfaction, perhaps that increase could have been twice the amount.

Once you have identified the repeat business, have a meeting with your salespeople to get their story on why others didn't repeat. In some cases, you may want to contact a few clients directly and discuss their experiences with your station.

As a rule of thumb, a business will lose 25 percent of its customers in a year. Some move away. Some choose to do business with the competition. It is one thing to lose 25 percent of your clients in a year. It is altogether something else to lose 70 percent of them in a month.

Act Soon

Supplies of good advertising prospects are limited. The sooner you identify the churned accounts, the sooner you can take action. Further, if your salespeople know you are paying attention to client retention, they will start to pay attention to it themselves.

Radio salespeople don't worry about profit — they worry about sales. That's the way it should be. Profit is management's responsibility. Salespeople worry about how much they are billing. Sales managers need to go a step further to determine how they are doing it. Measuring churn is the first step.

Chapter 18

Say The Magic Words
Sold Out
By Dwight Case

When you are sold out, the arguments stop about bonus spots and lower rates. And if you aren't sold out, then all you argue about is bonus spots and lower rates. So I am stumping for being sold out by using a grid card that has real negotiating power.

- By being sold out in advance, you can make your time more valuable and charge more to new clients.
- To sell out, you will need more salespeople.
- Being sold out does not mean you can offer less service to clients, but it tells you the least amount of revenue you will have on the books in the future.

Most grid cards have very few ways to move in the true sense of maximum inventory. Does your card have a different cost for Monday AM drive and Wednesday AM drive? How about noon on Monday and noon on Friday? Or PM drive on Tuesday and PM drive on Thursday? If you have a schedule for 24 commercials, are there 24 different costs or just two or three? Is there a spot cost as low as $10 and one as high as $150?

Make Your Time More Valuable

Your card should have all of these choices. Watching retailers price their goods based on the laws of supply and demand is a great way to get a rhythm here; being sold out 90 days in advance is a great way to make a profit.

Explain to your clients that their commercials will be pre-empted when another client pays more for that position. You want to be sold out way in advance so you can run up the value of your time by charging more to new clients.

To be sold out requires more salespeople than you now have. Some

Chapter 18
Say The Magic Words

folks have done it with nine to 12. My favorite number is 17.

Being sold out does not keep you from needing value-added stuff ... Does not keep you from calling on the client ... Does not keep you from playing the right tunes ... But it does tell you the smallest amount of revenue that you will have on your books three months from now.

Fill it up. Sell off the top of the card for everything.

The reaction of the staff and the customer to your station being sold out is a wonder to observe. You are indeed more important than you thought.

Chapter 19

Trick Questions
And Commonsense Answers
By Dave Gifford

Scenario: You're one of two finalists for a GM position you'd kill for. In your final interview, the station owner asks you the following seven questions, insisting that each be answered in 25 words or less:

1.) What is your philosophy of how to make money in this business?

2.) What is your philosophy for outselling the competition?

3.) What is your philosophy of how to sell Radio?

4.) What is your philosophy for increasing sales?

5.) What is your philosophy for increasing collections?

6.) What is your philosophy for developing a winning sales staff?

7.) What is the best advice you can give a new salesperson?

- Radio sales is not complicated. We just make it complicated.

- It's as simple as: More salespeople = more sales. More presentations = more sales. Bigger presentations = bigger sales.

- Too many sales managers are obsessed with how to implement certain tactics without first figuring out what to achieve and why.

- Strategies and tactics are born from a philosophy of Radio sales and certain guiding principles.

Now, before I give you my answers, close the book and — in 25 words or less to each question — write down your answers. No fair peeking ahead. On your marks, get set, go.

Back so soon? Let's see how your answers compare to mine.

1.) Sell more. Collect more. Save more.

2.) The station that sells the most advertisers wins. Sell more advertisers. Sell advertisers more.

3.) The station that helps the most advertisers wins. Selling is helping, and helping is closing. Selling is helping is closing. Teach. Help. Sell.

Chapter 19
Trick Questions

4.) More salespeople = more sales. More presentations = more sales. Bigger presentations = bigger sales.
5.) The station with the toughest collections policy collects first.
6.) No train, no gain. The best-trained sales staff wins.
7.) Ask and you get; don't and you won't.

Despite the simplicity of my responses, there is a point to this: I ask those questions of sales managers all the time, and the answers I get back — in 2,500 words or more, in most cases — are almost always self-incriminating. Why?

A Simple Strategy

The problem with most sales efforts is that they're formulated backward. Whereas a philosophy (a sales theory in this case) gives birth to certain guiding principles that lead to an overall strategy, too many sales managers are obsessed with the implementation of tactics without first figuring out what to achieve and why.

My answers are not so much answers as they are philosophies and guiding principles around which to build strategies and tactics.

Still, nothing is as simple as it sounds, right? Wrong. It is that simple. We just make it complicated.

Here is precisely how simple it really is: Multiply the number of salespeople by the number of presentations weekly per salesperson to figure the number of total staff presentations weekly. Multiply that figure by the staff closing ratio to get the number of total orders weekly. Multiply the weekly orders by the staff's average order in dollars to get the total weekly sales in dollars. Multiply that by 52 weeks and you will have your dollar figure for total yearly billing.

It's a pure numbers game. Based on a minimal closing ratio of only 20 percent per salesperson, consider my answer to question No. 4: More salespeople = more sales. With three more salespeople each giving 10 presentations weekly, your station would be giving 30 more presentations weekly. Multiply that by a 20 percent closing ratio per salesperson, and you would get six more orders weekly. Multiply that by an average order per salesperson of $1,000 to get a weekly billing of $6,000 more. Multiply that $6,000 by 52 weeks to get an additional yearly billing of $312,000.

Bigger + Better = More $

To illustrate my philosophy that more presentations = more sales, consider that if those three new salespeople each make 20 presentations per week, instead of 10, you could plan on $624,000, twice as much additional yearly billing.

If bigger presentations = more sales, ask each salesperson to increase their average order from $1,000 to $1,500 — in addition to making 20 presentations per week — and you would end up with $939,000, three times as much additional billing.

Radio is not rocket science. It's mostly common sense.

What strategies and tactics can you build from our seven answers? That's the hard part, but at least you have a solid foundation on which to build.

Chapter 20

Temporary Solutions, Permanent Problems
Is Your System What Ails You?
By Gina Gallagher

If you are having difficulty diagnosing a problem with new business development, perhaps you are treating a symptom and overlooking the root cause. Take your troubleshooting a step further and look at your system's infrastructure.

A solid sales infrastructure must manage the dual focus of immediate transactional business (avails) and the long-term developmental business (alternative sources of revenue). The challenge requires that you determine whether your current operation supports two distinctly different disciplines.

- Determine whether your current operation supports two distinctly different disciplines of transactional business and new business development.
- Examine whether solutions are addressing problems with your infrastructure, rather than symptoms of those problems.
- Identify possible side effects of those solutions.
- Empower your staff to explore solutions.

Treating The Symptoms

Most stations begin by developing a list of activities and standards designed to produce results; instead, this burdens a system that was designed for a different discipline and compromises both the transactional business and the new business effort. To encourage sales, they start by increasing the commission. They may hire a new business director, a consultant, provide video training or send salespeople to training seminars. While all of these actions are worthwhile, they only temporarily address the symptoms. Any initial spurt of activity eventually wanes.

Typically, we react to limited growth by pushing harder on the system. For example, management can use all kinds of incentives to push an overworked staff to sell more, but eventually this push will only back-

Chapter 20
Temporary Solutions, Permanent Problems

fire. You might notice some improvement, but the underlying problem gets worse. Let's examine some of the typical underlying problems that can limit your growth.

Problem: An overburdened sales staff. With the exception of an occasional "list janitor" on the staff, most salespeople work incredibly hard. They are overburdened with minutia, meetings and reports and are spending as little as two hours a day in actual face-to-face selling.

Question: How much of your reporting system is redundant? What is a more effective way to measure output? Should seasoned professionals have the same requirements as rookies? Can your communication be accomplished as effectively without a group meeting?

Suggestions:
- Empower your sales staff to think of efficient ways to communicate.
- Enable your staff to secure college interns to help with the time-consuming details. If your policy requires paying interns minimum wage and your budget cannot accommodate the expense, then suggest that the salesperson incur the expense with reimbursement when they hit new business goals.
- Scrutinize your reporting system and look for new ways to measure effectiveness, such as aiming for a certain number of new marketing proposals instead of a call report.

Defense And Denial

Problem: Overburdened support staff or lack of staff. Staff often complain that the station wants new business but fails to provide the resources to handle the extra work and won't acknowledge the legitimacy of the problem.

Question: How can we provide the staff with the resources they need? How can we acknowledge the legitimacy of the problem and not fix it? How can we fix the problem when we don't have the budget to hire additional support?

Suggestions:
- Provide the staff with laptop computers and let them pay for them with monthly installments or reimburse them when they achieve new business goals. They can compose the proposal on the laptop and give the disk to the sales assistant to clean up and print. They can keep a complete account profile and contact history that you can access and keep on a master file.
- Admit the limitation of the organization. When you react to the complaint defensively or with denial, you only set into motion another problem. The sales staff feels negated and powerless. The most common expression of this frustration is lower productivity. The reality is

that no matter how hard you try to create ways to bring about change, you must first deal with the underlying issues.

By exploring and recognizing the subtle dynamics of each part of your existing system, you can begin to create a holistic system that deals with causes rather than symptoms.

Chapter 21

Make Radio Tangible
Introducing The Radio Tear Sheet
By Michael B. Hesser

Does this complaint sound familiar? "I didn't hear my ads last week, and no one who came in mentioned that they heard them, either." Or, "My wife wants to know why I'm running ads on your station. All her friends listen to XXXX." Or, "Well, dear, I had a pretty good weekend sale but ... I don't know whether it was the paper, the shopper's guide, your station or the other station I ran my sale ads on."

Of course, anybody who's spent any time on the street has heard similar comments and questions. What do we do as Radio salespeople? We dance, wiggle and talk to save the account. However, there might just be a way to alleviate some of these frustrations, unnecessary hassles and defenses. I submit to you, for your own mental well-being ... the "Radio Tear Sheet."

- Skeptical advertisers complain that Radio's benefits are intangible.
- A Radio tear sheet provides tangible proof that an ad ran when scheduled and shows the advertiser the context in which the spot aired.
- Tape a new advertiser's spot off the air using a short-duration cassette.
- Include preceding songs, your announcer giving the time, the calls, etc., and then the commercials, your announcer again and the beginning of the next song.
- Take the recordings to your clients. It shows you care about their business.

For those of you unfamiliar with the newspaper tear sheet, it's proof of insertion — an advertiser's ad or the entire page is cut from the paper. The tear sheet is usually delivered to the advertiser before the paper comes out or immediately after, and it's included with the bill. If requested, send your Radio bills out with affidavits of performance and a copy of the script.

Chapter 21
Make Radio Tangible

Get It On Tape

Now let's hear it again: Radio is a tough sell because it isn't tangible. So let's make it tangible — especially to the newer advertisers.

First, you need inexpensive, short-duration cassette blanks. Second, print up cassette labels with your call letters, logo, phone number, etc. These tapes will become your Radio tear sheets.

For example, Baxter's Shoe Store wants to run a pre-Easter sale. Baxter breaks down and agrees to add your station to the media mix of four quarter-page newspaper ads and a heavy two weeks on another Radio station in town. Baxter is finally giving you a chance because you've put together a couple of "real cute spots," so you get 20 spots a week for two weeks.

Now's the time to leave nothing to chance; don't allow the advertiser to make the comments we discussed earlier. When Baxter's commercials are scheduled to run, you should be ready with your cassette blank to record — off the air — one minute before his scheduled cluster until after the stop set and into the next minute or so.

Hand Over The Evidence

Your recording should include the end of one of your station's "killer" songs, your announcer giving the time, the calls, etc., and then the commercials — Pepsi, Baxter's Shoe Store and the local Ford dealer. After the commercials: your announcer again and the beginning of the next great song. Baxter's intangible purchase is now tangible. But we're not done yet.

Drop in to see Baxter, bringing along the cassette, a player and a copy of the days and times his commercials are scheduled to run. Play his tape (remember it's only approximately three minutes long). Ask Baxter if he has a cassette player at home. If he doesn't, let him use yours for the evening so he can play his commercial for the wife and kids. They'll love it.

What has the Radio tear sheet accomplished? It has made the intangible tangible. It showed Baxter that he's in good company. It allowed him to hear one of his 40 commercials as it played to your listening audience. It gave Baxter pride of ownership and showed that you care enough about him and his business to take that extra step.

If your clients aren't happy after all of the above — blame it, as always, on your traffic director.

Chapter 22

Trial By Fire
Learning What No Seminar Can Teach
By Dr. Philip J. LeNoble

How have you trained your sales team? Videotape? Hotel seminar? RAB? Outside sales training? In-house? What are the results? Would you believe that, after all this time, your sales team is not doing anything different? That's right ... nothing different. Oh, each sales associate will tell you they got some very good ideas and how happy they were you gave them the exercise and that they look forward to using the material, but it does not go any further.

No video tool, hotel seminar, outside sales training or industry-sponsored training can take the place of continuous, in-field training application. If you want to teach piano, golf or swimming, you have to put the student in front of the instrument, hand them a club or get them in the water. The sooner your sales reps get in front of the customer, the better they will perform.

- No video, seminar, outside sales training or industry-sponsored training can take the place of in-field training.
- The sooner your sales reps get in front of the customer, the better they will perform.
- In the field, the sales manager can observe and coach new sales reps as they learn the product and the station's presentation.
- New reps cannot learn about the customer's business while sitting in hiding, in the station on the telephone.
- The sales rep and the sales manager prosper when they're in the field together.

Observe And Coach

Advertising schedules do not come out of the computer or in a seminar. They come from meeting the client where they conduct business. There has never been a simulation, a classroom model, a customized

Chapter 22
Trial By Fire

program, a videotape or hotel seminar that can produce the same rewards as in-field learning directed by the sales manager (who should always accompany the new rep into the field). There is no better training to follow the classroom than the sales manager and the new associate being in the field together, applying what was taught. In the field, the sales manager can observe and coach new sales reps as they learn the product and the station's presentation.

A new salesperson learns best by listening to "live" customers while working with the sales manager or a rep, who can pass on valuable clues and tips. If a sales manager sends the new hire into the field with a successful sales associate — not to make service calls such as picking up copy, but rather to make real presentations — the new employee can learn about the station's selling strategies. Additionally, when the new sales associate is exposed to the work habits of a successful salesperson and how much money they can make, the new rep will chomp at the bit to get out there.

The new sales rep has to get comfortable asking for an appointment, asking for information regarding the prospects, the business, asking for input, referrals or a start date. They cannot do this while sitting in front of a video or listening to a seminar.

Getting In Their Face

The most important part of the in-field sales training process is learning about the customer's business. New reps cannot learn while in the station on the telephone. They have to get in the customer's face. In-the-face client calls assure the sales rep and the manager that every positive technique is being applied to gain the competitive edge.

A sales manager can never meet these training opportunities while chasing paper in the office. In the field, a sales manager can demystify the sales training effort. Work with them shoulder-to-shoulder in the field and you'll never have a dissatisfied employee when they perceive you are doing everything personally to help them. When the new sales reps feel good about what you have taught them in the field, they will want to fly solo.

Think of the incredible turnover in Radio. The Standard Rate and Data Spot Radio Guide publishes close to 700 pages a month with all the changes. Currently, Radio gets a little more than 7.6 percent of the total advertising pie. Before you tell your new sales reps you cannot get out into the field with them because you are too busy chasing cost per point, consider this: The future holds and the present gives. Being where you need to be is the best way to get where you want to go. The new sales rep and the sales manager prosper when they're in the field together.

Chapter 23

Is Your Station Burned Out?
Or Is It Just Your Light Bulbs?
By Chris Lytle

A Kentucky sales manager recently mentioned that his owner can always tell when a Radio station is doing poorly. He counts the burned-out light bulbs in every station he visits. The more burned-out bulbs, the worse the station is doing. If people at the station are overlooking minor problems like light bulbs, he reasons, you can be sure there are far more serious problems that they aren't addressing.

That's not to say that, if there's a burned-out light bulb in your office right now, your Radio station is going to go dark. It's just something to think about.

- Burned-out light bulbs are a symptom that something else is wrong in a Radio station.
- If you have a Radio station that you don't want your clients to see, you may not be maximizing your sales.
- Clients need to visit your facility to understand they are buying a Radio station and not just Radio spots.
- People at the station need to meet the people they serve, and clients need to meet the people who serve them.

Get Clients On Your Turf

I suggest that salespeople give clients a tour of the Radio station. And, every time, this suggestion is greeted by uncomfortable giggles from one or two groups of salespeople. They look at each other and roll their eyes as if to say: "We can't bring clients in there!"

I have been in several stations in the past few months that I would not want clients to see, either. However, there are two good reasons to get the clients on your turf. The first is because it is your turf, and they will not be interrupted by phone calls, employees, customers or competing salespeople. The second reason is to show them they are buying a Radio station and not just Radio spots.

Chapter 23
Is Your Station Burned Out?

Your salespeople can make calls all day in their best clothes and their shined shoes. Yet, pretty soon clients will become detached from what they are really buying. Arranging for clients to meet the people in the office — the announcers and the production people — is good for the clients as well as for the people in the station. They, too, get detached from the people they are servicing. Meeting clients will help them remember that they aren't just "cranking out another log" or cutting another commercial.

Show Off Station Sophistication

It's good for clients to see that your station has computers, satellite dishes, transmitters, consoles and other sophisticated equipment that we take for granted. It's good for them to go into the studio and see their name on the log and their commercial in the cart machine (or whatever machine you're using these days). It's reassuring to see the station because they can't see their commercial — and neither can anybody else. Showing people the sophisticated delivery system is a way for you to make Radio more valuable to them.

You might even end your tour at the "prize closet" and select a CD or a poster for your clients' kids so they will be heroes when they get home at night.

There is another advantage to using station tours as a sales strategy. Agencies try to take the personality out of Radio by negotiating the best cost per point for their client. Dozens of media buyers in your area buy hundreds of thousands of dollars in advertising. How many of them have been inside your Radio station? When will you invite them?

Have you ever met a media buyer who didn't have a birthday? One of the toughest negotiators I've ever met — a media buyer — would find a way to justify your station if you attended her birthday party.

If Walls Could Talk

When you do summon the courage to invite the client in, remember that top sales managers and salespeople pay attention to the little details. After you've finished reading this article, get up from your desk, walk outside your door and look back in your office. Ask yourself these questions:

• What does the way my office looks right now communicate to my sales staff and to my clients about the kind of person I am?

• What assumptions might they make about me from the condition of my desk?

• Would I feel comfortable doing business here or following the person who occupies this office?

The burned-out light bulb could be a symptom that someone in ownership or management is burned out. It may signal to employees and clients alike that other things are being overlooked.

It could be that you just haven't had anybody worth impressing come to the Radio station for a while. In that case, it's easy. Set up tours with at least one media buyer and two direct clients for next week. Then you'll have to vacuum the floors, clean the coffee mugs in the sink and ... change the light bulbs.

Chapter 24

A Niche In Time:
The Mass Marketing Era Is Over
By Godfrey and Ashley Herweg

The era of mass marketing is swiftly coming to an end. The 18th-century Industrial Revolution that thrived on mass-production lines for mass markets is obsolete. The head-count mentality of newspaper circulation figures and television gross rating points is going down the same drain. Cost-per-point pricing and other mass marketing measurements that embrace 12-plus numbers make no sense in the new era of niche marketing.

Giants Respond

The response of mass-marketing giants to niche marketing might help put this revolution in perspective. Venerable companies like General Motors, U.S. Steel and IBM are downsizing or slowly going out of business. Consumers are demanding more choices, and some of those demands are being met. In the early 1940s, Granddaddy Coca-Cola sold one type of cola in a six-ounce green bottle. Today, Coke sells as many as 26 different sizes and flavors of soft drink products. Coke is trying to meet the demands of various market niches.

Hallmark greeting cards now markets more than 1,200 different cards just for Mother's Day. Mom can receive greetings from children,

- The era of mass marketing is coming to an end, and the era of niche marketing has begun.
- Cutting-edge companies learn their customers' smallest needs and cater to their customers' perceptions.
- In Radio, the local stations' advantage will be their ability to offer local niche marketing information.
- Learn what your listeners want, then present this information to your advertiser.
- Demonstrate highly targeted Radio formats tied to intelligent niche marketing plans that use your Radio station.

Chapter 24
A Niche In Time

spouses, parents and pets. And Mother's Day greetings are now given to a Mom-to-be, an Aunt Who's Like A Mom, a Stepmom, a Stepmom-in-law, a Grandmom, even a Great-Grandmom. In today's marketplace, Mom has many niches.

Cutting-edge companies now seek ways to learn their customers' smallest needs and cater to their customers' perceptions. The Japanese combined robotics and computer technology to create the ultimate niche marketing scenario. When a customer walks into a bicycle shop, the salesperson takes detailed notes about the customer's needs (a Customer Needs Analysis). The customer's specifications are fed into a computer that reaches the factory in nanoseconds. In less than 48 hours, the computerized production line builds the custom bicycle. To establish the value of a customized bicycle, however, the company takes two weeks to deliver the order.

The Local Advantage

Today, markets like Boston and Orlando have up to four different Radio stations in a group targeting four different niches. In the near future, when satellite and cable Radio deliver hundreds of potential choices, these niches will become even more narrowly defined. The local stations' advantage will be their ability to offer local niche marketing information that may come down to block-by-block research. Niche marketing opportunities are limited only to your imagination. At WHLP Baltimore, the niche is defined by jobs. The AM station devotes 16 hours a day to reading listings for jobs and airs features on job-seeking skills.

Know Your Niche

Your advertiser wants to know if your 18-to-24 Radio station can sell their flowers to sweethearts. Or sell their flowers for funerals, if your niche is 65-to-74.

A niche (the French word "nest") is simply a small corner of your market which you target and super-serve. You don't just focus on a narrow demographic. You study that demographic with a microscope. You learn what your listeners want ... better than anybody else in your coverage area. Then you present this information to your advertiser.

Conduct research ... about the client's customers, not about the client's goods and services. Go to your client's business. Question customers. Your objective is to find out not only what is selling but which products and services delight customers and why. Now, develop an advertising program to promote these niche opportunities.

Next, demonstrate the spending power of your niche format. Include the lifestyles, market share of population, income and geography of your niche audience as it pertains to your client's products and services. Demonstrate highly targeted Radio formats tied to intelligent niche marketing plans that use your Radio station. Show your clients how to capitalize on what their customers want. The stations that offer their clients the best marketing solutions will get the advertisers.

Chapter 25

The Nurturing Edge
The Five Basics Of Customer Service
By Mimi Donaldson

The future economic success of Radio lies in our ability to compete in a global marketplace, even if yours is a single stand-alone station. Audiences and advertisers have choices from around the world. That is why thousands of stations are focusing on customer service. Trainers are being hired; classes are being taught.

In my heterogeneous customer service classes, women seem to have an edge in grasping the rapport-building service mentality. I think it's because women discovered long ago that people often don't buy on the basis of need, and people don't sell based on their breadth of product knowledge. People buy people; the buyer buys you.

Women also grasp more easily the go-the-extra-step-service approach, as they've had to go many extra steps to achieve quasi-equal footing with men in the workplace. We are nurturers, and the concept of taking care of the customer comes naturally.

The basics of my customer service training, however, are not gender-based. I train customer service representatives whose main job is to answer complaints and calm agitated customers. They handle huge volumes of calls and must be compassionate while being efficient. Most people think this is an impossible task. It's not.

Here are the five basics steps of servicing a customer with grace and control:

1.) *Be clear on your purpose.* What do you want the customer to do,

> - **Taking care to give excellent customer service will give you the competitive edge.**
> - **Be clear on your purpose.**
> - **Be appropriate.**
> - **Know your "hot buttons."**
> - **Push the "pause button."**
> - **Give six-second empathy.**

Chapter 25
The Nurturing Edge

think or feel after your communication with them?

In the "do" column, we might list: Pay, renew, expand the order, tell friends to buy, not call my boss, never again call to complain.

In the "think" column: Think we're an excellent company, and I'm a capable, intelligent, professional person; think our product is worth the investment.

In the "feel" column: Feel taken care of, feel they're in capable hands, feel satisfied and confident in their decision to buy, feel trust in our company and in me.

When people are clear on their purpose and write it down in their own words, their focus improves. It's also the necessary basic to provide focus for the next four steps.

2.) *Be appropriate.* Appropriate is my favorite word in the English language. The dictionary definition is "proper, fit, suited to a given purpose." In *I Ching, the Book of Changes*, a source of oracular wisdom in Chinese philosophy for 3,000 years, a most important concept is Lu, which means "conduct." An excerpt: "One's purpose will be achieved if one behaves with decorum. Pleasant manners succeed even with irritable people." To the customer service representative, it means that every word and action must be suited to the purpose they defined in Step 1.

Logic prevails as people start examining their behavior. If your purpose is that this customer come back, would you be rude to prove a point? Of course not. If your purpose is having the customer think your company is professional, would you answer their query as to the whereabouts of a salesperson: "Oh, she's around here somewhere — we never know where she is." Ridiculous. These comments defeat your purpose. They're not suited to your given purpose, so they're not appropriate.

But how do you stop these sentences before they come out of your mouth?

3.) *Know your "hot buttons" and don't get sucked in.* Certain words or phrases used by customers push your buttons. Example: "What are you gals doing over there anyway? ... It's your fault ... Let me speak to someone who knows something ... You must have lost my payment ... Why is your product so expensive?"

Be aware of these traps. Make a list, read it over, desensitize yourself, so the next time you hear one of them, you do not have to lash back with a defensive remark, or a "yeah, but ..." Instead, you can ...

4.) *Push the "pause button" to gain control.* Our "pause button" separates us from the animals.

Some customers you know act like stimulus-response mechanisms. Their upsets are consistent and predictable. But your reaction doesn't have to be. When you are aware of your hot buttons, and one gets pushed, you can pause — very briefly — and choose the appropriate response.

5.) *Give the customer six-second empathy*. Using empathy is demonstrating with words that you understand what the customer is saying and how they are feeling. It is a statement that is calming, comforting, positive and specific, and a good one takes only six seconds. "I understand how frustrating it is not to get the information when you want it." Six seconds. "I understand how easy it is to get impatient with that machine." Six seconds.

A sincerely empathetic statement can defuse a hostile customer. It also gives you time to think of the response you can make which will satisfy your customer (i.e., achieve your purpose) while staying within the boundaries of your company's policy.

Chapter 26

Turn Today's 'No' Into Tomorrow's 'Yes'
Do's And Don'ts
By John Fellows

When confronted with "no," many salespeople doggedly press on, as they've been taught to do, hoping to overcome objections and turn the tide. Other salespeople get downright ornery, believing that a show of force will straighten potential clients out. Some politely thank prospects and suggest that they would welcome another opportunity in the future.

As a salesperson who's tried them all, I can tell you which approach delivers the most long-term benefits. How about you? Frankly, with all the talk at sales seminars about handling rejection and the everyday nature of declined proposals in sales, you'd think salespeople would quickly learn to handle "no" gracefully. But most don't.

Here's a short list of sales do's and don'ts to earn future business from today's "no." Pass it along to your salespeople.

- Don't let your ego spoil the potential for future business. Resist telling your prospect how you really feel.
- Don't question the buyer's judgment, even indirectly.
- Do encourage your prospects to consider your station in the future.
- Do send a follow-up thank you note.
- Do remember that "no" means "not now," not "never." Follow the course of pleasant, positive persistence.
- Do ask permission to come back.

Don't Burn Your Bridges

• Don't let your ego spoil the potential for future business. It's natural after investing considerable time and energy in a presentation to want to tell your prospects how you really feel about their decision not

Chapter 26
Turn Today's 'No' Into Tomorrow's 'Yes'

---

to go with your station. Don't. Tell your boss, your spouse, your therapist ... anybody but the prospect.

- Don't question the buyer's judgment, even indirectly. "You bought what?" Salespeople use variations on that line everyday, often without realizing it. In the eyes of the prospect, you're really saying: "Man, are you stupid." Not many buyers warm up to that line of reasoning.

- Don't get arrogant or huffy. One of the most common salesperson comebacks is: "Your competitor uses our station." Most business people aren't in love with their competitors, so the endorsement of a competitor is generally not a compelling reason to buy your station.

- Don't say anything you could regret later. "I'll never let you buy my station for less than X, you can be sure of that!" Lines like that can be prophetic, regardless of the salesperson's future efforts. Alternate sources, usually at better prices, are available to buyers of just about any product or service.

Amazing Grace

- Do encourage your prospects to consider your station in the future. "If there's ever something you think I may be able to do for you, I'd welcome the opportunity to help out." Keep the doors open. A decision today may be altered or reversed in the future, and a graceful acceptance will be remembered.

- Do send a follow-up thank you note. A "thank you" after being turned down? Absolutely. Since so few salespeople do it, it's one of the most powerful ways to demonstrate your maturity as a salesperson, and your interest in being reconsidered. Tom Hopkins offers some good samples of thank you notes in his book *How To Master The Art Of Selling*.

- Do remember one of the tenets of selling: "No" means "not now," not "never." Some people take this as a license to be a nuisance. Others follow the course of pleasant, positive persistence, knowing full well that cultivating and maintaining a positive image in the eyes of their prospects is one of the keys to business growth.

- Do ask permission to come back. Try your own version of this line: "Would it be OK for me to drop by from time to time when I have something I think you may have an interest in?" It works great to keep your prospects' doors open to you. Just live up to the promise. Only contact the prospect when you have something of potential interest.

All of us want more customers at one time or another, some more often than others. All of us hate losing "the big one." Using these simple, commonsense ideas can help us all earn future business from today's "no."

Chapter 27

To Kill A Vendor Department
Seven Deadly Sins To Avoid
By Val Maki

Here are seven points to remember what not to do if you want your vendor department to succeed.

1.) *Appoint a part-time vendor director.* There's no better way to communicate to the staff a revenue-generating position's lack of importance than to make the effort part time. Putting a part-time person in the position of leading the sales staff in business development will show the staff the department is not an integral part of your sales structure. This would essentially say to them: "Your business development efforts should be full time, and here's your part-time resource. Now, go get 'em."

Full Time Or No Time

Keep in mind that the vendor department can take on many roles under many titles which have a lot more to do with generating revenue than just co-op and vendor. It would be better not to have a special department at all if it can't be full time. Instead, a station could rely on account executive category specialists/trainers or outside help.

2.) *Pay a vendor director commission/override on vendor sales, but take that percentage out of the salesperson's commission.*

At first glance, it looks like a way to help pay for the department, but

> - Part-time leadership? It would be better not to have a special department at all if it can't be full time.
>
> - Split commissions? A successful business development department where one person is both a resource and teammate is always structured non-competitively.
>
> - Immediate results? It's better to hire the best people possible and let them find their own business development strengths.
>
> - "This can't work!" You accept defeat the minute you believe something can't be done.

Chapter 27
To Kill A Vendor Department

it will appear that you have ulterior motives and could cause pain. A successful business development department where one person is hired as a resource and teammate to the staff in selling is always structured non-competitively. Staff members should be thinking more about the business at hand than worrying about what they might lose by using a station resource.

3.) *Expect results immediately ... like within 90 days.* The vendor director's pay should not be tied to this time frame, either, in terms of guarantees and dropping of guarantees. It's better to reduce the risk factor in the office, hire the best people possible, be positive and let them find their own business development strengths. Let them make a mistake trying something new. This all takes time and is well worth the investment.

Excuses, Excuses

4.) *Allow yourself and the staff to use the excuse: "This market is different. Those types of programs don't work here."*

You accept defeat the minute you believe something can't be done. It is true that some staffs will gravitate toward one type of business development over another. Just remember two things when it comes to vendor sales:

First, in every marketplace there are manufacturers, retailers and service companies trying to sell more product and increase market share. This dynamic is a common denominator that makes every market an opportunity for the vendor program.

Second, just when we think something is not possible, someone comes along and does it.

5.) *Let veterans overlook vendor sales in lieu of "more important" pursuits ... thereby relegating the vendor department's business development to the retail (sometimes rookie) staff.* Why wouldn't anyone put their best-trained salespeople against their business development efforts? They could set the example and provide the experience and training for the less-trained of the staff.

Let Creativity Rule

6.) *Limit the types of programs the staff can sell which you will credit as being vendor.*

Train them all, then let creativity take over. New business is new business. An AE will use the training to increase spending from existing advertisers.

7.) *Don't involve other departments in the pursuit of vendor business and, by all means, don't communicate with them.*

Obviously, all departments need to know the vendor departmental structure, objectives and needs. Likewise, the other departments should communicate their own needs. Involve everyone, but especially the promotions department.

Chapter 28

Client Science
Building Rapport With The Three Vs
By Dr. Sharon Crain

In Radio, our competitors can match our product or our service, but our real advantage can be in the quality of the relationships we build with our customers. So it makes good business sense to study this particular science.

In contrast to our personal lives, where we may have the luxury of time to build rapport and empathy with those we like, our time with customers and prospects is often limited. However, awareness of a few basic elements can make a big difference.

> - The quality of client relationships can be your best advantage over the competition.
> - Building rapport requires awareness of the Three Vs of communicating: visual, vocal and verbal.
> - Especially during phone contacts, monitor your speaking voice to ensure an authoritative vocal image.
> - Be careful not to send an incongruent message with conflicting styles.

True Or False

Below are some true and false questions to test your awareness of these elements of building relationships. Mark your answers. Then, after reading the article, come back to find out how well you did.

1.) Immediate rapport between people is a natural phenomenon that is difficult to create. T/F

2.) Most new prospects remember more of what we say than how we sound when we say it. T/F

3.) Most men's voices carry less credibility than women's voices. T/F

4.) Similarity is the most important element in building rapport quickly. T/F

5.) How fast we talk should be determined by the importance of our message. T/F

Chapter 28
Client Science

Visual, Vocal And Verbal

Most of us have experienced meeting someone and immediately "connecting" with them. Significant research over the last decade has found that rapport and empathy between people is based on specific elements which we can create.

Underlying these elements is the awareness that as humans we have three major communication channels through which we influence others. These are known as "The Three Vs" — visual, vocal and verbal.

Studies have indicated a difference in the amount of impact among the three when we meet a new prospect, for example. Our visual impact will account for about 55 percent for a first-time prospect, vocal will account for 38 percent and verbal, only about 7 percent.

When contacting our prospects by phone, the 55 percent visual is obviously lost and is replaced by vocal. We then build a visual image from the vocal. Who hasn't had the experience of picturing the newly hired DJ as gorgeous because of their voice, only to find a totally different image when we drop into the studio to check them out?

Since vocal is the major channel on the phone, we need to monitor the sound of our voices for both credibility and rapport. Women's voices tend to carry less natural credibility and authority than do men's lower-pitched voices. Because of this, it is to our advantage to develop the awareness and skills to speak with intonation and inflection patterns that signal credibility.

Another critical element of the "Three Vs" of impact is that all three of our channels must carry the same message. For example, if our vocal lacks credibility but the words (verbal) sound authoritative, the overall message is incongruent and draws a negative reaction. The result is that we don't build the necessary rapport, and the prospect or client lacks confidence in us.

This same principle of incongruence applies when we meet customers or prospects in person and our visual impact becomes paramount. To express enthusiasm, we want to make a positive facial expression, project vocal vitality and use motivating verbiage. If one of our channels sends a different message, we are in danger of damaging our credibility.

Create Rapport Quickly

Prospects and customers feel comfortable with us when we create similarity with them. We can accomplish this with the "Three Vs."

In the visual mode, we create similarity by mirroring our customer's basic presence. Specifically, we would assume the customer's stance or sitting position and then generally reflect their movements and level of gestures. If this sounds absurd to you, test it on a casual contact such as a bank teller. Notice the immediate comfort level between the two of you. Mirroring takes practice and should be used subtly.

Vocally, the most important element is rate of speech. Think for a moment, if you are a "fast-talker," doesn't it drive you bonkers to converse with a slow-speaking prospect? Your prospect feels the same way about you. With a little concentration and practice, adjusting your rate of speech to match another's is quite easy — and you have the world to practice on. Remember also that rate of speech is more noticeable on the phone, since visual impact is lacking.

The verbal element of building rapport is known as matching predicates. This means we notice whether our customer speaks in seeing, hearing or feeling terms. A person who thinks in visual terms will say: "I see what you mean ... Can you see what I'm saying?" The prospects who rely primarily on their listening sense will say: "I hear what you're saying." The kinesthetic, or feeling type, will say: "I feel we should do it this way."

If a client says to us: "I feel you handled this poorly" and we say: "I see what you mean," we create a mismatch. To match predicates and build rapport, we would say: "I understand why you feel that way."

A colleague who travels extensively suggests a side benefit of becoming proficient at quickly building relationships. Before her flights, she approaches the ticket agent with her coach-class ticket and matches and mirrors the behavioral style of the agent. Then, with a high level of sincerity and enthusiasm, she will say: "If you have any seats available in first class, I would appreciate being considered."

Four out of five times when there are first-class seats available, she flies first class. This is a practical and beneficial way to prove to yourself that these skills really do work.

Chapter 29

Dialing For Dollars
Get The Appointment, Then Make The Sale
By Pam Lontos

The song *The Wanderer* is a great theme for salespeople who roam the streets daily trying to sell Radio time with no appointments. "I can't get an appointment on the phone" and "I'm no good on the phone" are common complaints. However, the telephone can be a salesperson's best organizer and timesaver. With proper techniques, appointments are easy to make and can increase your sales dramatically.

It is more professional to set appointments. Agency buyers and retailers are busy people. They don't want salespeople just dropping in on them with an attitude of: "Drop what you're doing. I'm here to sell you advertising." It shows no respect for the client's time. And it's hard to sell someone you've annoyed.

- The telephone can be a salesperson's best time-saver and organizer.
- It is more professional to set appointments before trying to make a sale.
- Try to reach the top decision-maker first.
- You are more likely to get through by having authority and confidence in your voice.
- Don't offer too much information about your station over the phone. Get the appointment first.

Setting an appointment by phone improves your chances of getting the sale. This is because the clients have allotted the time to listen to you. Also, clients are more receptive because you have shown respect for their time by phoning ahead.

Get To The Right Person

1.) Ask for the owner, president or senior manager. Don't ask for the person in charge of advertising, or you'll often get someone who can't make a decision on their own. This can lead to months of wasted time.

Chapter 29
Dialing For Dollars

2.) Start at the top. If your first meeting is with a person who is not a decision-maker, it is difficult to go over that person's head without antagonizing them. If your first meeting is with the top person, however, you can always go back to them if you don't get results with the subordinate.

3.) Use the name of the person at the top. If the top person, such as the president, refers you to the advertising director, let that person know that the president (using his/her name) told you to call.

4.) Get past the receptionist. Top decision-makers use their receptionists to screen calls. Since receptionists generally work set hours and decision-makers don't, you will have a better chance of reaching the boss directly if you call early (7:30-8:30 a.m.) or late (5:30-7 p.m.).

5.) Find out if the decision-maker is in before you ask to speak to that person. This makes it hard for anyone else to screen your call by claiming that the person you wish to speak with is out.

6.) Have confidence and energy in your voice. You must sound like someone of authority who should be put through.

7.) Use first names. With accounts who are hard to reach, use your first name and their first name to sound like a friend. Clients reached using this method often become the best long-term clients.

8.) Use assured, commanding words. Don't say: "May I please speak to Mr. Jones?" Instead, say: "I need to talk to Bill."

9.) Get the secretary's first name and use it. If you make the secretary your friend, your messages are more likely to get through.

10.) Don't leave messages. Clients don't say: "Oh, boy, another salesperson. I'll call." They simply don't call back.

11.) Don't immediately introduce yourself as a Radio salesperson. The receptionist will block your call; or, if your message does get through, the client will say: "I don't want to talk to a salesperson." Then all your future calls are blocked.

12.) To avoid leaving a message, say: "I'll call back. Thank you." Get the decision-maker on the phone first and create a good rapport before you tell them you are in sales.

13.) If you have to leave a message, leave your name and number only, with an interesting message that will arouse curiosity.

14.) Make cold calls in person if you can't get to the decision-maker after several attempts. Don't just walk in and say: "I want to talk to you right now about Radio advertising." Instead, introduce yourself. Say: "I know you're a busy person. I only came by to set up an appointment at a future time." Often the decision-maker will see you right then.

Organize Your Calls

1.) Set a definite time to call each day — preferably from 9 to 10 a.m.

2.) Plan to call at least 30 people. Make a tally mark for each person you call so you know when you reach 30. Thirty calls should result in reaching 15 people, which should yield eight appointments. Make 30 calls daily and you will get appointments. With practice, your ratio of appointments to calls should improve.

3.) Once you start your 30 calls, don't stop until all are done. These calls go fast if you don't stop to talk or go for coffee in between. When you get an appointment, dial another number immediately.

4.) Always set appointments for the next day. If you always have appointments, then you can't make excuses and spend the day at your desk when you're tired or unmotivated. After you start talking to the prospect, you'll get energized.

5.) Call back at the end of the day to reach all the people you were unable to get in the morning. Call between 4 and 5 p.m.

6.) Don't call during your peak selling time, 10 a.m. to 4 p.m. Be in front of clients then.

Getting The Appointment

The purpose of the phone call to the prospect is to sell an appointment, not to sell your station. Withhold as much information as possible about your station so the prospect can't put you off with objections on the phone. Obviously, you will often have to give some information, but the less you give, the better. Objections are easier to deal with face-to-face than over the phone. Besides, if you give away information on the phone, it eliminates the reason to see you. Create curiosity.

Chapter 30

Getting Stores To Play Along
Tips To A Good Kickoff
By Kathryn L. Maguire

Here is the scenario. You put together the best promotion ever for a manufacturer. He sells the idea to one of his biggest accounts, a retailer with lots of locations. He is paying you big bucks for this campaign, and you deserve it because in addition to the 100 spots he is purchasing over a four-week period, you are providing a trip to give away in his account's stores, printing registration boxes and tear pads, and going with him to make the presentation to his buyer contact at the account.

Finally, the big day arrives — the promotion kickoff. You decide to drop by a couple of the stores to see how the displays look with the registration boxes and tear pads. You walk into the first store and head for the display. But wait ... where is the display? It's not up yet. Uh, oh, you can't find the registration box. You ask the store manager where the display and registration materials are, and he looks at you blankly. "What display?" he asks. You go to another store. Same story.

What happened? You were there when the manufacturer sold the promotion to the buyer. The buyer agreed that the displays and registration materials would be no problem.

- Send sizzle letters to outline store responsibilities. Hold a store manager meeting.
- Hold a display contest to give stores incentive to participate. Have a registration ballot contest.
- Call the managers to personally tell them about the campaign. Send sizzle tapes.
- Alert the stores that a Mystery Shopper will be checking out the displays.
- Have store managers sign and return their information sheets; offer a prize.
- Visit stores on the day the campaign kicks off.
- Hold a signage contest.

Chapter 30
Getting Stores To Play Along

Do It Differently

Don't take it personally, but do something differently in your next promotion. Make the stores aware of and accountable for the promotion, because quite often communication from the buying office to the store goes haywire. Here are some ways to fix the problem:

Send sizzle letters. These are little one-sheets sent to the store or department managers to alert them to the promotion and outline their responsibilities.

Hold a store manager meeting. At the time this promotion is sold to the buyer, ask if you can hold a kickoff meeting with the store managers to get them excited about (and aware of) the campaign.

Run a display contest. One way to make sure displays get up is to offer an incentive. Award the store or department manager with the most attractive display. Restaurant trades with theater or concert tickets go over big for these contests.

Have a registration ballot contest. This will ensure that registration boxes are put out and tear pads are always on hand for customers. Give a prize to the store manager who collects the most registration ballots.

Call the store managers. Call them to personally tell them about the upcoming campaign. If you have interns, this is a great job for them.

Mail sizzle tapes. Have your production person put together a sizzle letter on tape. Add on the commercial. Use sound effects! Store managers are not accustomed to getting cassette tapes in the mail. Curiosity alone will ensure they are played.

Reward Participation

Mystery shopper. Send out a sizzle letter first, telling store managers that the mystery shopper will be dropping by certain stores to check out the displays or other promotion elements. If everything is up, the store manager automatically wins something.

Mail out return cards. This is a sizzle letter with a twist. At the bottom of your information sheet, you ask the store manager to sign it and send it back to the station for a prize. Great prizes: movie tickets, sports tickets, station paraphernalia, etc.

Visit the stores. Drop by some stores the day the campaign kicks off to see how things are going. If there's no sign of your promotion, at least you are early enough in to the campaign to fix it before it's too late.

Run a signage contest. Ask each store manager to snap a picture of the in-store signage or other promotion materials and send it to you for a prize. Although it is unlikely that every store will do this, you will have great pictures to use when you follow up with your manufacturer and retailer.

The biggest reason you should take responsibility for what happens in-store is the result for the manufacturer. He is always going to be looking at the bottom line — sales. If the promotion goes as planned, then his sales should increase; otherwise, he wouldn't have agreed to do the promotion. You can sit at your desk at the end of the campaign and point fingers at what everyone else was supposed to do, but when it comes time to sell the next program to your manufacturer, he might not be as open to your ideas. Use these tricks to make your promotions work for everyone, and show the good service that comes with the campaigns your clients buy.

Chapter 31

Becoming Self-ful
Assertive Communication At Work
By Mimi Donaldson

Do you find it difficult to express what you want and need to the boss? Are you unable to respond when you think you should? Are you frustrated by your powerlessness in some day-to-day interactions?

The art of confidently and comfortably expressing your wants and needs without hurting or being hurt is a crucial skill. Few of us learned the art of assertive communication from our families. As a result, we are ill-prepared to meet the challenges of the workplace, where people need to get results through other people. Priorities compete for attention, and the "squeaky wheel" (often, the overly aggressive person) gets the grease, especially in an ego-driven environment such as Radio.

> - Develop a "self-ful" attitude — the art of being confidently assertive.
> - You can learn how to say no and have people thank you for it with a three-step method.
> - When you speak up to the boss, advise him/her of your priorities, just like you do with everyone else.

The most important issues in life are about needing or not needing the people we work with. It's about confronting, "assuming," standing one's ground and, most of all, about courage. We have to make a choice — over and over again. We have to choose between telling the truth to someone who needs to hear it, or keeping the truth tucked away and unsaid. We must choose between being comfortable and safe, or risking discomfort and even the loss of some of our perceived popularity. We also choose, every day, between our hot-button response ("You can take this job and shove it!") or the appropriate response suited to our long-term purpose.

Patience And Hot Buttons

Maturity is a measurement of patience: how long you can put off

Chapter 31
Becoming Self-ful

immediate gratification. We all know that you must put off a hot-button response ("I'm just sure ... does it look like I have four hands?") for a long-term result. Being patient involves self-confidence. There are three different behaviors to choose from:

1.) *Selfish:* Since that time long ago when we whiningly started a sentence with "I want" and our mother called us "selfish," we have been fighting that label. We've gotten it confused with "aggressive," "pushy" — worse terms when applied to women.

2.) *Selfless:* This is the non-assertive person who avoids conflict, at all costs. They wimp out of calmly expressing needs and wants. This person is not confident of his/her rights as an employee and as a human being. These rights are: to be treated with respect, to be listened to and taken seriously, to have and express feelings and opinions, to ask for what you want and to get what you pay for (how many of you have paid for a bad haircut — and given a tip?). When we act selfless, we become a natural victim for every aggressor. They ignore our subtle signals of martyrdom, and attend to their own priorities at our expense. People who ask: "Got a minute?" end up taking half an hour because we wimp out of saying no.

3.) *Self-ful:* This is a word I created. It doesn't mean "full of yourself." It stands for a person confident enough of their rights to be assertive: to ask for what they need and want without hurting other people. This takes skill and practice. It is the art of saying "no" to people and having them thank you for it. Don't think it's possible? Assertive, "self-ful" people use a three-step action method. Here's an example:

Tom knocks on top of your cubicle partition, leans in and asks: "Got a minute?" Instead of glancing at your watch and saying "OK" with a martyred sigh, you look up and analyze the request. You see his lower lip trembling and his eyes filling with tears. You know he wants to talk about his divorce — again — and you have a report to finish. You recognize this will not be a 60-second interruption, no matter what he said. You resist the reflexive "hot buttons" response ("In your dreams, pal") because you depend on Tom in your job. A rapport with him is a priority for you. Take the following three steps:

1.) *Acknowledge:* Use six-second empathy to tell him you understand how he feels and what he wants. "Tom, you look upset — it looks like you need to talk." This calms him, because now he doesn't have to work to make you understand. You have said, in essence: "I understand your priority — and it's important."

2.) *Advise:* Let him know your priority — calmly, "self-fully." You start out: "Tom, here's the situation. I have a report to finish for the network vice president, and it's due in half an hour." You have understood his need, and now you're asking him to understand yours. Many people, when told of your priority, will back off. But not Tom. That's why there's a third step.

3.) *Accept or Alter:* Accept the interruption with time limits ("I can give you five minutes") or suggest an alternative or option ("I'll come to your cubicle when I've finished the report").

What About The Boss?

With peers, you have the "alter" option; Tom will actually thank you and go away happy. With the boss, your best option is almost always to accept. The boss's priorities are your priorities — it's in the job description. However, don't leave out the second step. Always advise the boss of your activities and priorities. Sometimes, you are keeping them informed and they're grateful. And sometimes they want you to do it all anyway. This is when negotiation comes into play. But never skip step two. That's the "self-ful" step.

Being self-ful allows you to speak up and say what is important to you. It even allows you to correct the boss when you notice an error. Better sooner than later. Remember — bosses hate surprises.

Chapter 32

Radio's Little Instruction Book
So Simple, So True
By Chris Lytle

Life's Little Instruction Book is a big best seller. It contains page after page of truisms (one to a page). It's advice that requires no further explanation.

I wasn't in the mood to write another book. Perhaps these thoughts on our business will get you through a month of sales meetings.

Copy & Copywriting
- In advertising, we do our best work for people we like.
- The headline is the ad for the ad.
- The worst thing a client can say to you is: "You're the advertising expert. Work something up for me."
- "Attention Gardeners" is a better headline than "Now that spring is in the air ..."
- When you write copy, you have to focus on your prospect's prospects' problems.
- Guilt sells.
- Radio advertising doesn't work. You are simply testing the copy and offer against the target audience.
- Read each piece of copy and ask yourself: "If it were my money, would I approve this script?"
- "God's gift to Radio is that people are born without earlids." — Tony Schwartz

Selling & Closing ...
- The four stages of learning to sell Radio:
1.) You don't know you don't know.
2.) You know you don't know.
3.) You know you know. And finally ...

Chapter 32
Radio's Little Instruction Book

4.) You forget what you know and just do it.
(At this point you are usually promoted to sales management. Sales managers have forgotten what they know and must train people who don't know that they don't know.)

- In school, 93 percent was an A. In Radio sales, 25 percent is an A. In advertising, a 2 percent response is considered "genius." You can fail a lot and still make a lot of money.
- Resistance is the reason for the existence of sales.
- Objection prevention is a far more important skill than objection handling.
- The most important selling skill is asking questions.
- If you have more than 50 clients on your account list, it is not really an account list. It is a hunting license.
- You don't have to trick people into buying Radio. The best closing line is: "This is right for you. Let's do it."
- When all three commercials in the stop set are yours, it's called a Power Set.
- Sales at the newspaper go up for three weeks when Arbitron comes out.
- Your clients get better when you get better.
- No one single sale matters.
- When you control the focus of the call, you control the call.
- Selling Radio is different, but your market isn't.
- Media kits have never sold one second of air time. When a client asks you for a media kit, it is a stall, not a buying signal. When a client asks for your media kit, act shocked. Say: "We don't send out off-the-shelf media kits anymore. We customize everything." Then ask: "Who is the client, and what do you want in your media kit?"
- As long as you're in business, you're in sales.

Management & Motivation

- There are no "people problems." Every business problem is a management problem — and most management problems are systems problems.
- If you have weak salespeople, you need stronger systems. Either you need systems to help you find and hire better people, or you need selling systems that weaker people can work.
- Only in cases of extreme deprivation does money serve as a motivator.
- Your top biller may no longer be your best salesperson.
- There are two kinds of managers: growing and obsolete.

Secrets Of Success

• The biggest investment you will ever make is your career. Most people think their house or their retirement fund is the biggest investment they'll ever make. But your career is the money machine that will fund your house and retirement.

• It takes three to five years to establish yourself in business and 10 years to master a profession.

• "There are two kinds of people: drifters and deciders." — George Odiorne

• There are three secrets of success:
1.) You've got to know what you're doing.
2.) You've got to know you know what you're doing.
3.) You've got to be known for what you know.

• It takes three years to maximize an account list. The first year of your career, the learning curve is higher than the earning curve. Once you've learned enough to be of service to people, your earning curve quickly rises.

• "Problems of belief are more critical than problems of technique." — Dr. Jeffrey Lant

• There are more great sales jobs than great salespeople.

Only George Burns can survive for very long on one-liners. However, there is enough common sense advice here to help you coach and develop people for the next few weeks.

Chapter 33

Take Me Seriously Or Take Me Dancing
What Does A Woman's First Impression Say?
By Dr. Sharon Crain

The 30-second first impression rule states that others make major decisions about us within the first 30 seconds of seeing us. At a first meeting, they don't yet know how smart we are or how well we can perform, but they do have a lasting impression of our visual impact.

Most women acknowledge that our ongoing challenge is to establish the credibility that is often automatically awarded to the male gender. Since 55 percent of our impact is visual, we want to be certain that this impression propels us into the "take me seriously" category.

- Because the first 30 seconds often make or break a first impression, visual impact is critical.
- Women often must work to establish credibility that is often automatically given to men.
- Five basic personal elements measure visual image: hair color, hairstyle, skin color, height and weight.
- Women should choose their professional dress based on the seriousness or softness of their total visual image.

Credible Costuming

Most of us have had a lifetime of training in choosing clothes for our social lives. We choose these clothes based on what colors, styles, fabrics and textures are most flattering. In our professional lives, though, we have a different purpose: to enhance our credibility and make the correct visual statement for our economic purpose. We rarely get as much guidance in crafting this image.

I have found the "costuming approach" to be most helpful in providing some objectivity on our total image. To use this technique, you must jump out of your own head and imagine yourself as a casting director responsible for costuming yourself for your specific role as a

Chapter 33
Take Me Seriously Or Take Me Dancing

woman in Radio.

Personal Basics

The first category of self-evaluation involves our personal basics. These elements significantly affect the way clothes appear on us. The five personal basics are: hair color, hairstyle, skin color, height and weight.

Each element of our personal basics sends off a separate psychological cue to others. Each of these cues is on a continuum which runs from "take me seriously" on one end to "take me dancing" on the other. Rate yourself below on the following elements.

Hair Color: The darker your hair color, the more authoritarian your cue. The lighter your hair color, the softer your look. A more medium hue places you toward the middle of the continuum.

Hairstyle: This is a critical cue to your overall image. The more simple and tailored your cut and style, the more serious your image. Hair length is a major factor: Shorter hair sends more tailored cues, while longer hair (especially longer than shoulder length and curly or wavy styles) projects a softer look.

Skin Color: The darker your skin color, the more serious your look. Black or Mediterranean skin coloring with dark hair projects a strong authoritarian image. The opposite is obviously true — fair skin with light hair sends a "take me dancing" look.

Height: If you are tall, your overall impression will be on the serious side. A shorter woman projects a softer look. Since height is an unchangeable physical factor, this in no way means we are doomed to a less-powerful role. Some of the most dynamic women I know are short.

Weight: A lean stature sends off visual cues of control that translate into "take me seriously," while a fuller figure emits a less-powerful look.

Total Image

Now, total the five personal basic dimensions to determine if you have more on the "take me seriously" or the "take me dancing" side. Be aware of the degree of each of the dimensions. If you are very strong on several, those can overshadow the more moderate dimensions.

The purpose of this exercise is to evaluate your "before costuming" profile. This information gives you a tool with which to select the clothes that will project the look you want to achieve for specific situations.

For example, when it is important to appear super-credible, a woman who has a basic soft look might dress in a tailored navy suit to achieve her goal. This very same suit on a woman who already has a basic serious look could cause her to appear cold or unapproachable.

Keeping that 30-second rule in mind, ask yourself what you want your first impression to be, then select your hairstyle and professional wardrobe accordingly. Do you want that potential employer, subordinate or client to take you seriously ... or take you dancing?

Chapter 34

Follow Up Or Foul Up?
Service After The Sale
By Nancy Friedman

O f all the people who sell products or services to you, how many take the time to follow up after the sale?

I've trained businesses all over the country and have concluded that there's something important missing. It's called service after the sale. You get courteous, friendly treatment before you've spent your money. But after you make your purchase, it's as if you've ceased to exist.

What an unpleasant way to do business and what a costly mistake. Attracting a new customer costs five times as much as keeping a current one,

- Service after the sale is an important element missing in many business relationships.
- Attracting a new customer costs five times as much as keeping a current one.
- Two-thirds of advertisers who leave a station do so because they were treated with indifference.
- A strong follow-up plan can give your station the edge over competitors.

according to the Service Edge, a consulting company in Minneapolis. Once you have made the investment to obtain a new customer, why not keep him? The customer already knows you, your company and how well you deliver. The second sale should be the easy one. Your hottest prospects are right in your own sold files.

Even if a sale was completed flawlessly, you could still lose that advertiser by failing to take a personal interest in his or her satisfaction. Two-thirds of advertisers who leave a station do so because they were treated with indifference, according to the Service Edge. *The Wall Street Journal* reports that the average American company will lose 10 percent to 30 percent of its customers this year. Most leave because they are dissatisfied with the service.

Chapter 34
Follow Up Or Foul Up?

Get A Leg Up With Follow-Up

Your station's competitors have comparable prices and similar features. Every time you lose an advertiser to them, they become a little stronger and you become a little weaker. A strong follow-up program may be your only competitive tool to maintain healthy, long-lasting customer relationships.

How do you get advertisers to come back or send referrals? By planning a well-developed, deliberate strategy. The goal of a follow-up plan is to give the advertiser a reason to do business again.

You can make your follow-up programs as creative or as ingenious as you wish. Customize your program to your advertisers. Consider it part of the marketing mix and then allocate funds to accomplish your objectives.

The simplest way to follow up is with a thank-you letter, note or post card just to say "hello." Christmas or holiday cards are an appropriate gesture but may get lost in the shuffle. Send a card any other time during the year and it will stand out. Remember, a thank-you card should be just that. Use direct mail or other means for selling.

If writing letters or notes doesn't appeal to you, pick up the phone and call the advertiser. Cover the same points as in a follow-up letter. Is there an interesting article in a current journal your advertiser may want to read? Drop it in the mail. Are there tickets available for an upcoming ball game? Plan how your follow-up program can generate goodwill.

Send a box of candy, a calendar, a ballpoint pen, mug or other useful gift for no reason. Tastefully display your station's call letters, telephone number and logo on the item.

For special clients, use your imagination. One of the most unusual gifts I ever received was from a printing company that created its own wine labels and delivered personalized bottles of chardonnay at Christmas. It was a delightful way to publicize its new four-color printing capabilities.

Most of these techniques cost very little and are easy to implement. Compare the cost of thank-you notes to taking out an ad in your local paper!

Boil And Bubble

I've discussed these follow-up ideas with many salespeople. Every once in a while, one of them says to me: "Yes — but I'm hesitant to call. What if they have a complaint?"

If that advertiser has a problem, it won't fix itself. By contacting an unhappy advertiser, you have an opportunity to solve the problem, sometimes easily. If you don't call, the advertiser will likely be gone forever. Even worse, you have allowed the person to "boil and bubble" and tell others about his complaints.

Disputing with advertisers over small issues is counterproductive when you consider the value of a lifetime business relationship. Studies show that it can take up to 12 good customer service experiences to overcome a bad one.

Most people won't fight. They'll switch — to your competition. If you have unhappy advertisers out there, be a problem-solver. The choice is yours. Follow up ... or foul up.

Chapter 35

Use Software To Maximize Sales
A Practical Guide
By John Fellows

Today's software is easier to use than ever before and will help you earn more money — if you let it.

Contact Management

Contact management software (CM), and the related telemarketing programs and Personal Information Managers (PIMs), are the software backbone for street-level sales reps. Notable: ACT! for Windows, DOS and HP95LX; Commence; Lotus Agenda; TeleMagic.

Use your contact manager to:

1.) Remind yourself to send notes to seasonal clients/prospects during their off-season, and create personalized form letters.

2.) Target categories of prospects, send personalized prospecting letters and schedule timely follow-up appointment telemarketing calls.

3.) Design templates for appointment confirmation letters, thank-you notes, proposals, script forms, prospecting letters, etc.

4.) Autodial when phoning.

5.) Print your appointment calendar and task lists daily.

6.) Print sales reports. Managers and reps can have the entire contact history for every client/prospect instantly available for quick printout, neatly presented in a standard or custom format.

7.) Remind yourself of important dates: client birthdays, anniversaries, sale events, etc.

> - Contact managers keep you on track, efficient and on time.
> - Word processors give your documents professional polish.
> - Graphics programs add visual impact to presentations.
> - Ratings programs can favorably position your station, whether it's big or small, if you use a customer-focused approach.

Chapter 35
Use Software To Maximize Sales

8.) Create professional "custom" proposals in minutes with a standard format.

9.) Prompt you to call to confirm appointments.

10.) Track your mileage and expenses, and print out a monthly, quarterly and/or yearly report for taxes.

Word Processing

Word processing (WP) programs do lots of things typewriters never could do: print in multiple fonts and sizes, quickly rearrange words on a page and insert illustrations, charts and graphs, to name a few. Most contact managers include a word processing module, but a standalone word processor and a decent printer can create even more tasteful proposals. Notable: Microsoft Word, WordPerfect, AmiPro, Framemaker.

Use your word processor to:

1.) Create personalized mass mailings, i.e., station/rep lists, commercial lists.

2.) Design and print custom packages for specific clients, client categories, seasonal offers, etc.

3.) Design and print custom additions to your standard media kit.

4.) Design proposal presentation pieces.

Graphics

Graphics or presentation graphics programs are great for making bullet lists, charts, graphs to use in media kits, presentations, proposals, etc. Your data can go on paper, transparencies, slides or on your monitor screen. Best of all, you can string multiple charts together to create a screen show. Excellent when used in combination with a well-thought-out proposal from your WP or CM. Notables: Harvard Graphics, Corel Draw, Lotus Freelance, Microsoft Powerpoint.

Use your graphics program to:

1.) Compare an advertiser's target demo, income or gender cells to your station/format profile by demo, income or gender cell. Great way to clearly show a perfect match between the audience your station reaches and the customers your advertiser is trying to attract.

2.) Make bullet lists, graphs and charts for proposals. Create a standard (but easily customized) station overview for clients/prospects. Output it to overhead transparencies, slides, paper or as a screen show on your desktop or notebook computer. Used in combination with a presentation binder and a custom proposal, these screen shows can really "wow!" a prospect and help you close more sales.

Ratings Programs

Ratings programs help advertisers get a numerical picture of your station's strengths in the marketplace. Simply show your station's strengths relative to the advertiser's target, not relative to the other stations in the market. All you need to do is toss out the "bigger is better" mentality along with the "rankers" and stock up on a customer-focused marketing mentality and the "audience profiles" available from the ratings programs. Notable: The NAB publishes an excellent Directory of Radio Computer Software Suppliers.

Use your ratings program to:

1.) Become a single-source Radio resource. Create a sales presentation binder containing all your station propaganda plus ratings printouts of: audience profile for your station and every other station in the market, hour-by-hour listening for your station and all others in the market, reach & frequency reports showing the R&F for every station in the market for selected numbers of ads and the number of ads required to attain selected frequencies (both reports for selected demos) rankers by daypart for selected demos, rank projections for all the stations in the market by selected demos and dayparts based on your current rate. This is good for showing your apples-to-apples price comparison to the other stations.

2.) Run a reach & frequency report for the specific schedule in your proposal and/or that the customer is buying. Adds tangibility to your schedule.

Chapter 36

Courtesy Sells
Old-Fashioned But Basic Business Sense
By Jack M. Rattigan

Do you like being treated special? Are you impressed when people remember your name? Does a friendly voice on the phone make you feel good? If so, you appreciate courtesy.

What about your station? Do your people practice old-fashioned courtesy? Or has the dog-eat-dog attitude penetrated your staff to the point where they have forgotten the essence of being nice to people? Being nice can have dramatic rewards. When the rest of the business world isn't practicing niceness, your personnel can stand out and shine as real professionals. That goes for on-air personalities, salespeople, the business office and, of course, the receptionist.

- Old-fashioned courtesy is the basis of professionalism.
- Rudeness or indifference to clients can lose business for your station.
- Often, clients hesitate to tell management when staff is unpleasant or impolite.
- "The customer is always right" should be every business's credo.

Rudeness Loses

Do you ever stop to think how many listeners we turn away with a rude on-air personality at a remote or personal appearance, a "what do you want?" receptionist or a "take it or leave it" salesperson?

If you don't think it is happening at your station, if you don't think a disagreeable person has a negative effect on your station, allow me to relate an unpleasant incident I experienced not long ago. I was checking into a hotel in a city where I had acquired a new client. When the hotel manager heard that I was the station's consultant, he wanted to talk to me. At first, he told me how much he liked the station and what a great job it did for his business. Then he changed his tone of voice. "You have

Chapter 36
Courtesy Sells

your work cut out for yourself, buddy," he said. He told me that the salesperson who calls on him is the rudest person he deals with. He said: "She called me the other day to tell me that the station had a great promotion coming up. She informed me that she would call back in an hour and I better make up my mind or she would give the promotion to another hotel. You tell them if she calls me again, I'm finished there. Now you know how I feel, buddy." (I was hoping I would be his buddy before I left town. More important, I was hoping he would be the station's buddy before I left town.) I informed him that I was certain that the station management was not aware of the circumstances. I was right.

On further investigation, station management learned of other similar unhappy advertisers. The surprising truth was that these advertisers and former advertisers had good relations with station management but never told them of their unhappiness. Since I was the guy from out of town, it was easier to unload on me. After all, as one person told me: "You're here to straighten this place out. You better start with teaching them courtesy." (Actually, I wasn't there to straighten out the place. The station was doing very well. I was retained to hold sales development sessions. They just wanted an objective analysis of their operation.)

Professional Politeness

Unfortunately, I have discovered similar conditions in other markets. Advertisers hesitate to complain to the station. Sometimes when they do tell the station, it is too late. The client cancels or doesn't renew but never gives the real reason. Generally, they say: "I didn't want to cause any trouble." The real trouble is that the client-station relationship is destroyed and no one at the station knows why.

A little old-fashioned courtesy, rudimentary politeness — genuine professionalism — could prevent these problems. We must teach our people that the customer (listener and advertiser) deserves respect, that we are here to serve them and help improve their business. It is sad that we have to remind our people that the customer pays the bills and puts the money in their paychecks.

Tom Peters, the "Guru of Service," refers to Stew Leonard's Super Market in Norwalk, Connecticut, as an outstanding example of customer service. I made a special trip to Stew Leonard's some months ago to see for myself. His credo is carved in stone at the market entrance: Rule No. 1: The customer is always right. Rule No. 2: If the customer is wrong, go back to Rule No. 1.

Let's keep that in mind and remember that a little courtesy can turn into a lot of success.

Chapter 37

If Not CPP, What?
A Discussion On Radio Pricing
By Bill Burton

Cost per point and cost per thousand have been discussed vigorously at our recent Detroit Media Directors Council meetings.

The Detroit Radio Advertising Group set up the council with the primary objectives of: a better understanding between buyer and seller, and finding more-effective ways to move the client's products with the use of the Radio medium.

Most media directors had very strong opinions on CPP, but all were open to better alternatives. Unfortunately, no one had a better idea. Some of the points put forth were:

> - Buyers feel that there must be a means of substantiating their buys for clients.
> - Demo pricing could pose even bigger problems for Radio, giving the buyer a "club" for beating down prices.
> - Some researchers say that effective frequency is more important than CPP in reflecting the value of Radio.

- "You want to sell me your Radio station because it gets results. If I don't get results, do I get my money back?" — Bob Mitchell, senior vice president and group media director for Lintas-Campbell Ewald.

- "Should we be buying Radio as a direct response medium?" — Randy Schroeder, executive vice president and media director for Campbell-Mithun-Esty.

- "We must have some type of measuring tool for our clients. We have to substantiate what we're buying." — Ron Fredrick, senior vice president and group broadcast director for J. Walter Thompson.

- "Why not cost per cume, which more accurately reflects a station's total audience?" — John Cravens, VP/GM WHYT, Detroit.

Chapter 37
If Not CPP, What?

Radio Bargains?

The meeting also opened discussion in other Radio pricing areas. Michael Browner, general director for media operations with General Motors, said: "Radio on a cost per commercial can be very appealing, but on a cost per thousand is no bargain compared to television."

Mitchell said that demo pricing is a far bigger problem than CPP. He stated that most Radio stations price their station to get on every buy. Example: A station with younger demos, to be effective on a 25-54 buy, may lower its rate to $100. The following week, within the same agency, they may want to double the price when it hits in the heart of their primary demos. That, Mitchell said, gives the astute buyer a "club" to beat down your price tag in any future negotiations. Stations respond by saying: "If you're going to evaluate me by CPP, which is driven by my demo rating, I have no alternative but to demo price."

Bob Mancini, senior vice president and director of media services at J. Walter Thompson, said if he were a GM of a Radio station, he would develop a two- to three-year plan, narrowing his pricing so it would be more realistic for both buyer and seller. He believes that this would create more confidence that the product is fairly priced.

Frequency Is Fundamental

A number of different opinions have been put forth both pro and con on OES-Optimum Effective Scheduling, which was laid out in the excellent book by Steve Marx and Pierre Bouvard. Optimum Effective Scheduling to me means effective frequency. In my opinion, frequency is fundamental to selling anything.

Charlie Sislen, marketing & research vice president at my alma mater, Eastman Radio, has this input: "GRPs and CPPs are not the best way to buy Radio, because it does not take into account the balance of reach and frequency." Sislen feels that effective frequency — the actual number of times a consumer must receive a message to be motivated — is much more important.

Gerry Boehme, senior vice president and director of research for Katz Radio Group, agrees. Since reach and frequency are the two fundamental elements provided to an advertiser, he says it makes perfect sense to plan and buy based on those criteria.

Do You Have A Better Idea?

From my vantage point, one thing is certain: All advertising is going to have to become more accountable. We in Radio must do everything in our power to get results for the advertiser.

Chapter 38

Sales Warfare
Focus On Enemy (Competition) For Plan Of Attack
By Jack Trout

A war consists of a large number of small battles. While the overall strategy is critical, the actual outcome is determined by the tactical success of each battle. This is as true in business as it is in real warfare.

The Order Of Battle

In determining what tactic to deploy in each of these small battles, the field commander's first concern is the enemy's order of battle.

Each salesperson is a field commander representing his or her company at their respective accounts. Their job is to figure out what tactics to employ against their competitors to achieve an optimum share of the account's available business.

All too often, instead of studying the order of battle of their key competitor, the field commanders spend all their time studying the customers. Companies encourage this with sales training programs that teach how to figure out key influences or what best motivates these influences, and how to use the proper psychology to get the order.

The future belongs to sales organizations that become more competitor-oriented in their approach to dominant accounts. "Sales warfare" embodies that approach. It utilizes basic marketing warfare principles and treats each account as if it were an individual war of its own.

- Defensive sales principles are used to keep the competition out, not to maximize sales and profits.
- With offensive principles, businesses analyze the leading stations' strengths and weaknesses.
- Flanking is the tactic to use when you're getting only a modest amount of business from a customer.
- Guerrilla warfare is the "nothing to lose" tactic to use when you're getting almost no business from a prospect.

Chapter 38
Sales Warfare

The first step is to chart the account on a strategic square, which is divided into four principles: defensive, offensive, flanking and guerrilla.

Defensive Sales Principles

If your station has the largest share of an account's business, you are engaged in defensive warfare as far as that account is concerned.

Defensive Principle No. 1: The objective is to keep the competition out, not to maximize sales and profits. All too often, management pressures sales reps to make the numbers look as good as possible. Better to block and never let competition into the account, whatever the costs.

Defensive Principle No. 2: Never be afraid to sell against yourself. If you sense that your customer is sitting on a commitment that isn't up to what is now available from your station or from your competition, it's time to upgrade your account. This may mean helping your customer get rid of your older schedules and programs. It might be painful, but it's the only way to keep your competitors from exploiting a weakness.

Defensive Principle No. 3: Don't be shy about asking for help to block the competition. Just as a platoon leader has to "call up" artillery support, so a good sales manager has to be willing to call for help when a competitor launches an attack.

Offensive Sales Principles

The next quadrant on the strategic square is offensive sales warfare. This is when your competitor has the high ground and is getting the lion's share of the budget.

Offensive Principle No. 1: Analyze the competitor's strengths and weaknesses. Beyond the standard analysis of your competition's benefits, this will often require collecting some useful information at your account — with a certain degree of subtlety. Companies are generally forthcoming on what they like and don't like about other stations.

Offensive Principle No. 2: Find a weakness inherent in the leader's strength and focus your selling at that point.

Let's look at a non-Radio example. Say you're selling equipment to a big national account through a dealer chain. You analyze your bigger competitor's strengths and decide that it lies in their distribution channel. In other words, they have the best dealers in the country because they are committed to this channel; the dealers run the show and contact the client.

Approach this client on a hands-on, corporate basis. Your dealers would still service the business, but your management would call on them. You could invite the client to key trade shows where, unlike your competitor, they will get a heavy dose of tender loving corporate care.

Offensive Principle No. 3: Don't spread your efforts over too many features and benefits. "Selling the lineup" is fine when you're the lead station, but when you are attacking, a narrow focus is better. And the more features you try to sell, the less your chances of exploiting a weakness. Once you've established a breakthrough with a single feature, you're in a better position to sell others.

Flanking Sales Principles

In flanking sales warfare, you're not up against a single big competitor, but rather two or three big players who are getting most of the business from the account.

Flanking Principle No. 1: This is a tactic to use when you're getting only a modest amount of business from a client.

This is a small share game. Your objective is to increase your share by going where none of the leading stations is focusing attention. This is an opportunity for creative selling by using some imagination to figure out a unique angle.

Flanking Principle No. 2: Look for flanking opportunities in areas such as scheduling prime positions, quantities, sale terms and other non-traditional sales approaches.

Let's just say you are selling advertising to Jones Chevrolet, but purchases have stalled because their market share is declining and management wants to cut expenses. Consider an organizational flank: Look around the dealership for a department that hasn't been affected by the cuts, such as the service department. If it catches on, you'll have some business while the other stations are still waiting for things to improve.

Flanking Principle No. 3: Try to capture as much business as quickly as you can. The more business you get and the deeper you penetrate a new area, the more difficult it will be for your competitors to horn in.

Guerrilla Sales Principles

Guerrilla Principle No. 1: Guerrilla warfare is a "nothing to lose" tactic to use when you're getting almost no business from a prospect. Obviously, you have to determine whether you'll find happiness doing business with this company.

Guerrilla Principle No. 2: Try a number of "off the wall" sales approaches to get your foot in the door. It could be an unusual proposal or a courageous guarantee.

Guerrilla Principle No. 3: Focus on the most-productive approach. An example of this could be a non-traditional effort built around a narrow product focus rather than the whole store (i.e., the service department).

Be careful about suddenly attempting to broaden your efforts. You

Chapter 38
Sales Warfare

are the most vulnerable to a counterattack in the early stages of an effort. You don't want to threaten your competitors before you are strong enough at this new account to handle such a countermove.

Chapter 39

Leading A Turnaround
Some Suggestions
By Gary Fisher

Increased sales. Improved revenue shares. Up Arbitrends. Juiced department heads. Improved morale. A stable infrastructure. Good market buzz. Well-regarded hires. Cash flow increases. Picture-perfect income statements. Laughter in the halls. Food fights in the kitchen.

These are the unmistakable results of a successful Radio turnaround. There aren't many endeavors with as huge a financial and psychic payoff as turning a loser into a winner. Start-ups, move-ins and turnarounds are among the most passionate and challenging parts of the Radio business.

Just about any underdeveloped or underperforming station can benefit from the orderly application of turnaround management strategies. However, a successful turnaround requires a unique culture and managerial technique. Following are some points to ponder on the turnaround track:

Stake Out Your Niche — And Own It

1.) *Don't just target — segment and then hypertarget.* You can't build a house without a foundation, and the foundation to everything in a successful turnaround is reliable and actionable research. Only the audience knows what medicine an ailing station needs. Ask them, listen to them and obey them.

Research extensively to find out exactly what age-sex-cell, lifegroup and/or musical cluster you want to serve, and get ready to sliver-cast toward it. In busy markets right now, the AC audience is fragmenting into smaller particle markets. Niches are the rage in Radio; make sure your turnaround is based on owning one.

To successfully reassert itself in a busy market, an emerging station

Chapter 39
Leading A Turnaround

needs not only a viable format, but a different one. It must be distinctive enough to issue a promise and be the only station that can deliver on it. Once the chosen format is launched, the station's and company's commitment to it must be unwavering. These days, most major market stations operate with a brain-trust of group heads, consultants, national PDs and researchers, and all must be of one mind about the format's viability and shelf life.

2.) *A successful turnaround will usually come at someone else's expense.* It's important to estimate who will probably subsidize the turnaround and to what extent. Paint a target on someone's back and obsess about them daily. Most emerging stations do a better job when they have a target to aim for.

3.) *Sometimes you need to go back to the future.* Take in as much data (both objective and subjective) from the staff about the recent past as you can handle. The first weeks of any turnaround effort should be filled with individual staff meetings and debriefings. Talk to the existing staff, find out who they are, what they're all about and what they think needs to be done to turn things around. You'll find some gems as you mine for data, and the staff will feel more like helping you out.

Half of what everyone will tell you is probably true. The trick will be to distill the factual from the personal. It's important in any turnaround not to throw every baby out with the bath water. Sometimes you need to study what was working (when it was working) and possibly remodel after that to rekindle any equity left in the market. In the case of WNIC, this meant rededicating to the Harper and Gannon concept in morning drive, "Detroit's Nicest Rock" as the slogan or brand name, a warm and fuzzy family-values-based stationality and bringing well-known Pillowtalk host Alan Almond back to the party to handle nights.

4.) *Don't just plan to "get around" to fixing sales.* Start fixing it first to capitalize on the first signs of the turnaround. In a typical turnaround situation, the station's sales department is often suffering from tired blood. A history of not achieving budget, losing share of market and income erosion is bound to pull down confidence and morale. Given the urgent need for revenue and sales improvement in any turnaround situation, it's important to start upgrading the sales effort right away.

5.) *Start working in all possible ways to bring non-ratings-related demand to the station's inventory.* A key ingredient in a successful turnaround is momentum, or at least the perception of momentum. Make lots of direct retail sales and celebrate the orders internally. Commit totally to the Law of Massive Numbers — massive numbers of salespeople calling on massive numbers of accounts working in massive numbers of different developmental areas selling massive numbers of commercials to generate massive results. Nothing soaks up inventory at decent rates like retail sales, jingle sales, Yellow Pages sales, recruitment sales, prize catalog sales and the like. While duopoly and FM-FM combos will inevitably shrink the number of broadcast owners, it will do nothing to shrink the bottomless pool of Radio inventories, and turnarounds usually have plenty of inventory to burn off. Do anything and everything to press the existing inventory.

The Mission

6.) *Gradually and steadily raise the thresholds of acceptable performance in programming and sales.* This is far more complex than merely raising expectations via new sales budgets and ratings projections. It involves establishing a new vision for the station, a new culture of winning and showing how the most basic activities can contribute to winning. New levels of productivity and new measures of winning must be defined within the station's mission statement. The mission statement should be broken down into subgoals with a separate action plan for each goal.

7.) *Fortify the administrative and back-office ranks as soon as possible.* The station will need a smooth business infrastructure to capitalize on a successful turnaround. Lock it in early and let it percolate while you work toward the turnaround.

8.) *Do informal strategic planning with the department heads each month.* Continually update the station's mission statement and regularly compare where you are now with where you were, where you want to be and where you're budgeted to be. Analyze what is sure to be a fluid competitive environment and spotlight problems or opportunities as they become apparent. Continually update action plans for every subgoal in the station's turnaround plan.

9.) *Create strategies and action plans for every hill you don't own but want to in programming, sales and promotion.* Also, create defensive action plans for any hills you already own but find under attack. As an example, if your goal is to "own" vendor sales and to have the best vendor sales program and profile in your market, your strategy might include hiring an experienced vendor director, allocating a start-up budget, setting up hotel and printing trades, and joining the proper associations.

Getting Your Sea Legs

10.) *Keep communicating omni-directionally and aim to be ultra-accessible whenever possible.* Everyone must communicate their individual roles in helping to turn the station around. Every department head and staff member should understand how their performance influences other departments and the overall good. Department head meetings, state-of-the-sta-

Chapter 39
Leading A Turnaround

tion meetings, walk-a-mile-in-my-shoes cross-training sessions, all provide forums for celebrating wins and recognizing progress. Create an open-door policy, making sure people's needs for communication are known and workable. In the early stages of a turnaround, everyone's needs for productive access to the boss can often be overwhelming. But as the turnaround gels and progresses, the staff will quickly get its sea legs.

11.) *The one-minute manager is right-on for Radio.* Roam the station with an eye toward catching people doing things right. Put up communications boards, publish and celebrate small wins and successes, rave about one department to another, lock in get-togethers and burn up the trade as fast as possible. Wipe out classism; lock in cross-department thank-yous.

12.) *Be a catalyst for the department heads' successes — and then let them own those successes.* A station full of "owners" will outperform a station full of employees every time. The best way to build a management team of owners and then keep them "bought in" over the long haul is by re-entrepreneuring and decentralizing at the outset.

Hiring smart, managing loose, decentralizing quickly and praising often is the surefire way to keep things high-tech yet high-touch. All departments need mental preparedness and toughness to blast through the plateaus and resistance levels that all turnaround efforts encounter. Help your staff stay excited, enthusiastic and raring to go daily. Cheerlead the staff every day. Your optimistic, uplifting attitude must infect the staff and keep them pumped. An unwavering positive demeanor will be job one, especially during the rough periods.

13.) *After the turnaround is under way, stay obsessively close to the station's three constituencies: the listeners, the advertisers and the staffers.*

14.) *Manage people, not paper.* The uncertainties and delayed gratifications of a turnaround project require more one-on-one face time than at other types of stations. All three of the station's constituencies — listeners, advertisers and staff members — need highly individualized attention. Encourage meetings with clients instead of letters, grassroots and one-to-one direct mail rather than mass media, one-to-one huddles with staff members instead of memos.

15.) *Help your staff re-balance their lives long-term on behalf of the station.* Set station policies that allow department heads and other staff members to easily reconcile work/family issues. Make sure health care, medical, maternity and sick leave policies complement your "employees first" stance, so that they are not hidden obstacles to esprit de corps. If you've hired smart, you'll be able to manage loose.

While sticks, start-ups and turnarounds are somewhat unbankable and unsalable today, they still represent tremendous opportunities for present owners to re-equitize, re-create asset values and re-ignite cash flows ... the three qualities that propelled the lender-driven '80s and the three qualities that will help Radio get back to being a healthy, vibrant industry.

Chapter 40

Don't Fire The Copywriter Yet!
Solutions To Getting Great Copy ... On A Limited Budget
By Judy Carlough

Your sales projections were too optimistic. Expenses are running higher than expected. Revenue is slowing. Budget cuts are inevitable. So you:
  a. Fire the afternoon team and allow Malcolm, your deep-voiced engineer, to take the shift;
  b. Fire the business manager, give each department head his or her own checkbook and let them pay their bills;
  c. Replace your play-by-play announcer with your receptionist (hey, she's always been a big football fan);
  d. Fire the copywriter and let the account executives write commercials for your clients.

- Effective Radio copywriting is a craft that should be handled only by experts.
- The money a station might save on a copywriter's salary would eventually be lost — when clients see a lack of results from ineffective creative.
- When budget cuts are inevitable, there are other options instead of simply cutting the copywriting expense.

Every one of these solutions is absurd and would hurt your station's sound and profitability, yet only the first three are likely to induce cardiac arrhythmia in most general managers. Why?

Highly Illogical

Letting salespeople write copy is a white-haired Radio tradition that has survived illogically and, in surviving, has cheapened the image of our medium and the way our programming sounds. The practice was born in an era when there was no TV and few ad agencies had any noticeable expertise at the new art of writing broadcast copy. This practice may have worked adequately in a kinder and gentler era of limited media choices, little clutter and relatively unworldly listeners — but what

Chapter 40
Don't Fire The Copywriter Yet!

about today, when marketers are extremely sophisticated, and so are the audiences they target?

It's a formula for disaster. The art and craft of writing great Radio creative is no less difficult than playing pro baseball or painting a masterpiece. Can you imagine handing a palette of paint to an AE, pointing to a ceiling, offering her a few tips from Michelangelo, then saying: "Go for it"? The gap is profound between gathering the information and producing the words, sounds, voices and music that will transform into a distilled message to generate buying action by the audience. Only experts, never amateurs, can do it consistently. A number of amateurs can put words on paper, but few of them generate meaningful results. And isn't this the whole point? If the commercial doesn't generate a response, the schedule won't be renewed and (get ready for more arrhythmia) revenue will fall. Just try to picture the poor AE who wrote the copy going back to the client and trying to talk him back on the air.

It's a ridiculous situation, one that all our competitors seem able to avoid. Even in the smallest markets, the newspaper rep doesn't go to his hardware store client and say: "Thanks for the order, Maryann. Now, can I take a few minutes and sketch those hammers and drills for next week's ad?" When we allow non-professionals to write copy, we send the message that we don't respect our medium enough to require professional production. No wonder we aren't always taken as seriously as our competition.

The First Cut

So where's the solution? Where do you make the cut? The first step is to re-examine the budget (one more time) and cut something that truly is expendable or worthy of delay. If the cost of the copywriter's salary must be eliminated, there still are some ways to avoid the AE-do-it-yourself route:

- *Charge the Client:* Keep the copywriter but underwrite his or her salary by charging the client for that expert's services. Train your salespeople how to sell creative, the same way they sell sponsorships, features, sports and promotion packages. Put them to work at what they're good at. Make sure your creative product is first-rate, the same way you would any programming feature. Then sell it as a benefit to the advertiser by saying: "Our production department has won more awards and produced more results for advertisers than anyone in town." Then document it. But be careful, it could become a profit center — and how would you explain that to corporate?

- *Pool Stations:* Get a group of stations in your marketplace together to share the cost of a team of copywriters. Non-competing stations may make you most comfortable, but it makes sense to investigate similar demo formats; that way, the copy team can develop real expertise and familiarity with your audience. Don't worry about confidentiality; the team will honor it or be fired. Production can be done at individual stations on a rotating basis. Clients may be charged a nominal (or not-so-nominal) fee.

• *Restructure Commissions:* Did you poll your sales department and ask them if they wanted you to fire the copywriter? You might find that they're willing to give up a small percentage of commission to underwrite part or all of that salary, especially if it's only until revenue picks up again and the budget crisis ends.

• *Think Free-Lance:* Here's a novel concept: advertise! Run ads for free-lance copywriters and see who responds. For all you know, there are some veterans in your market with outstanding credentials who would work free-lance. Again, clients would pick up the cost. The free-lancers could work directly with the client and free up your AE to do more selling. Among other things, you'd save on employee benefits.

• *Explore Agencies:* If times are tough, agencies may be cutting back. There could be talented copywriters on the street. You might find someone who really knows automotive and could help you not only produce great commercials, but also close business. Hire on a free-lance basis, and charge the client accordingly.

• *Purchase A Copy Book:* Many universities and broadcast organizations such as the NAB and RAB have ad copy books. Professionally scripted spots also are available in co-op books.

• *Hire The Pros:* Did you know that some of the big Radio production guns have prepackaged campaigns available on a market exclusive basis? The cost to your client is extremely reasonable and would give them the best spots possible at a fraction of the customized price.

• *Scout Local Talent:* Comedy clubs, local theater and even colleges and universities might be a great resource. Comedians and actors often write as well as perform. Even if they don't write, they might be willing to work for a small fee to get exposure. Again, the client would pick up the cost.

• *Invest In A Great Sound Library:* Great production, using great talent, can sometimes (though not always) make up for mediocre writing.

• *Train Your People:* Train your salespeople how to convince clients that it's in their best long-term interest to invest in professionally produced commercials. They'll get results, they'll be committed to Radio and they can win awards — maybe a Mercury — and brag to their neighbors. Your sales team should know who the best independent producers are and what they charge. You might even negotiate a finder's fee.

Chapter 40
Don't Fire The Copywriter Yet!

All these suggestions require effort and education ... but it's worth it. The station that chooses the route to quick-fix savings by eliminating a copywriter's salary might be satisfied briefly, because the budget is balanced. But in a month or two or six, when business continues to fall, you'll see the visible results of the invisible cost of poorly conceived ads that didn't work, and clients who didn't renew. That cost has never been measured — and it's killing us.

Chapter 41

Know The Trades
Know Your Accounts
By Kathryn L. Maguire

To keep abreast of what manufacturers and retailers are talking about, read the trade magazines they read. *Supermarket News, Women's Wear Daily, Drug Store News* and *Computer Reseller News*, as well as periodicals such as *USA Today, Wall Street Journal, American Demographics* and *Brandweek*, are among the available resources. There is a category for whatever business you are most interested in pursuing. Just watch the trades to see what the next trend(s) will be. This will help you help your accounts, and you will be more credible on the calls.

Here is a peek at what is going on in the top vendor/retailer categories:

- Manufacturer and retail trade magazines are good resources for trends and issues in account categories your station is pursuing.
- Women are a growing target in the automotive category, with businesses catering to women and families.
- Nutrition is a big issue with food/grocery businesses, while environmental concerns are reflected in food and clothing trends.
- The growth of the home improvement industry is a reflection of the economy.

Women And Cars

Women are the target for new car manufacturers. *American Demographics* reports that women make up 49 percent of the new-car buyers and 23 percent of the new-truck buyers. Women have some influence on 80 percent of all new-car sales. This is changing the way vehicles are being built. Minivans and other large vehicles (designed for children) are hot. Aggressive car dealers are attracting women by offering "play areas" for kids and diaper-changing tables in the rest rooms.

Do-it-yourself sales are on the rise. Supermarkets and general merchandise stores sell automotive supplies, and auto supply stores

Chapter 41
Know The Trades

are more mass-consumer oriented.

Safety is a trend that extends from automobiles to parts to tires. Air bags, preventive maintenance and Michelin's "baby in the tire" TV ad are proof of that.

A Grocery List

In the food/grocery category, EDLP (every day low pricing) is the manufacturer trend that Procter & Gamble started in 1991. EDLP means lower prices on many key brands and streamlined trade promotion dollars.

Targeting products and promotions to ethnic groups, particularly Hispanic consumers, is a trend in this category and many others. This minority population is expected to be the largest one in 15 years.

Nutrition is also a hot trend in the food/grocery category, with FDA labeling requirements among the major issues. Low-fat and low-sodium products, fruit juices and waters are growing in popularity.

Environmentally safe (or green) products are also in demand. Consumers are recycling. Manufacturers are turning out "clear" beverages and laundry detergents for a healthier, cleaner or safer perception.

Microchips & Paint Chips

With new microchip technology dramatically improving the look of a video screen, we can purchase photo CDs, audio CDs, game CDs, computer program CDs, movie CDs — all either already on the market or on their way.

Computers are getting smaller in size and in price. Notebook computers fit our mobile working society and are popular with students, too. Competition for market share is fierce, and this always means avenues for Radio stations in search of new business opportunities.

Home improvement is another growing category. With stores like Home Depot moving into our markets, women represent 50 percent of the customer base. With home sales down nationally, home fix-up is on the rise. This category covers home furnishings, window treatments, floor coverings, wall coverings and paints, furniture, gardening and more.

Sneakers are big in the clothing category. There is a type for every mood: stepping, climbing, dancing, aerobics, cross-training, tennis, soccer. Clothing is more comfortable: loose-fitting jeans, colorful T-shirts and women's flat dress shoes.

This category is also beginning to reflect consumers' concern for the environment and other social causes. Naturally dyed clothing, natural fibers and "Made in U.S.A" are examples.

Chapter 42

Taking Control
The Systematic Approach To Career Stability
By Dr. Sharon Crain

As the uncertainty surrounding the '90s seems to increase daily, women in Radio management can bolster their sense of career stability by becoming more systematic. If your job should be caught in the downsizing landslide, there is a proven system when you step into a new position. This personal system significantly reduces the feeling of "overwhelm" that often accompanies fast change.

The system I recommend is based on two basic Total Quality Management (TQM) principles. The first speaks to increasing effectiveness; the second speaks to increasing efficiency.

> - A systematic approach to adapting to change uses the Total Quality Management principles of effectiveness and efficiency.
> - The first step is to define the client's Critical Success Factors, things which make them want to do business with your station.
> - The second step is to determine how to efficiently make those CSFs a reality.
> - By charting your steps toward that goal, you can regain control of your destiny.

Effectiveness is defined simply as doing the right things. This means managing priorities to direct your energy toward activities that produce the best results. In Radio customer service, effectiveness means meeting the needs of clients. Yet if your station's competition is also working to meet customer needs, you have to run harder just to stay in the game.

A wise strategy is to define a goal not to merely meet the needs of clients — but to delight them. Clearly, this goal would place your station well-ahead of the pack. How you accomplish this objective is where the TQM systematic approach comes into play.

Your first step is to identify the client's Critical Success Factors (CFS). Specifically, for each important client, identify those critical fac-

Chapter 42
Taking Control

tors that are necessary for them to want to do business with your station. As you identify those factors, you will be in an excellent position to know what factors are important for them to increase their business.

In Radio sales, we often think we know our clients' hot buttons, like price, for example. Yet when we delve into specifics, we find that price can be mitigated by a host of other factors. It is widely recognized, for example, that customers are often willing to pay more for increased time flexibility. Studies also show that price is not the major factor for customers switching companies; poor service is.

Gaining Control

In addition to the first challenge of identifying your client's CSFs, your second challenge is to efficiently make those CSFs a reality.

Serving customers according to their specific delights often requires restructuring your time and personal schedules. For example, the routine of scheduling appointments, prospecting and calling on clients may have to change dramatically. This often leads to uncomfortable, out-of-control feelings at first. You may feel your time is being used inefficiently, running helter-skelter from client to client attempting to "delight" them.

At this point, the second TQM principle of creating efficiency becomes important. First, chart the steps that you need to take to accomplish your goals. This systematic charting process will put you back in control of your time. By charting each activity, you can pinpoint unnecessary steps (held over from an earlier routine, perhaps) and remove them.

By doing this, you are both effective (delighting your clients) and efficient (achieving this end in the most productive way).

If the nature of the work place in the '90s continues to be chaotic, this systematic approach will help you personally excel in any position — in fact, in any field. You will have some major control over your destiny.

Chapter 43

The Real Reasons Clients Buy
They Have Little, Or Nothing, To Do With Your Rates
By Chris Lytle

New research by Learning International reveals that the top three reasons people buy have nothing to do with price, but rather relate directly to the quality of a sales force. Here are the three top reasons customers buy:
- Business expertise and image
- Dedication to the customer
- Account sensitivity and guidance

And yet, you still have clients who tell your salespeople: "Your rates are too high." Clients give us rate objections because they have trouble articulating the real reasons they aren't buying. Very few clients will tell you: "You are an inept, insensitive person dedicated only to selling time on your station to earn a commission and unwilling to offer me any expertise or guidance."

Instead, they let you down easy with: "Your rates are too high." The truth might be too painful.

> - New research shows that clients want expertise and guidance more than low price.
> - Sales force quality is the key to increased sales.
> - Sales training must be based on eliciting behaviors.
> - Shift your focus from selling your inventory to showing the client how to sell his or her inventory.

Researched Vs. War Stories

War stories are fine. Telling people: "This is how I sold such and such an account" can impart sales knowledge. However, being sensitive to research can help you conduct better sales training sessions. In training your troops, you are trying to impart specific knowledge, skills and attitudes to increase desired behavior in measurable ways. According to training expert Robert Pike: "A training program that doesn't change

Chapter 43
The Real Reasons Clients Buy

behavior is as useless as a parachute that opens on the first bounce."

You want salespeople to behave in a way that research shows the client wants them to behave. Train your sales reps to prepare for a call by looking into the industry they are calling on, to demonstrate more client sensitivity and to offer better guidance.

Here's a profitable example: A Charleston, S.C., sales rep discovered by reading an RAB Instant Background that 75 percent of the people who eat at a family restaurant decide on a restaurant within two hours of eating. Further study revealed that food quality is six times more important than price to the consumer. The rep recommended to the restaurateur that he advertise from 3:30 p.m. to 6:30 p.m., and promote the quality and freshness of the specials for that night. That behavior (making a recommendation based on the research) resulted in a 52-week order and increased sales for the restaurant.

Behavior Demands

To give clients what they want, you might have to change what you want from your reps. If you demand 10 calls a day, then expect some of them to be calls where salespeople are walking in unprepared.

"Anything coming down for me this week?" is the worst opening line in Radio sales today. But it's more than an opening line; it is behavior that does nothing to build business expertise and image, dedication to the customer or account sensitivity and guidance.

Here are five things to do now based on the research:

1.) Require salespeople to do some homework for important calls. (If a call isn't important, why are you making it?)

2.) Discuss a particular industry in a sales meeting and develop a creative strategy for the category.

3.) Coach salespeople before they make the call instead of after. Ask: "What are you going to present to them? What are you going to recommend they do?"

4.) Examine proposals for client focus vs. station focus. Are your people presenting solutions, or are they selling the station?

5.) Focus sales meetings on how the client can sell more of his or her inventory, instead of how you can sell more of yours. The harder you work to sell your clients' products and services, the easier it is to sell your station to them.

Clients want expertise and help. If your salespeople aren't providing that, then you can be sure clients will demand the lowest rate. Training your salespeople to behave like clients want them to behave is the profitable choice.

Chapter 44

The Furniture Boom
How Radio Reaches The Buying Generation
By Dr. Philip J. LeNoble

If your station is targeting the baby-boom generation, listeners who were born between 1946 and 1964, the furniture industry needs your station's consumers. NOW. Dr. Ken Dychtwald, who wrote *Age Wave*, says this group of 80 million buyers has a potential spendable income of $300 billion! Their spending will become the long-term capital growth of the furniture industry.

The furniture industry has long been marketing itself to the newspaper reader who, according to Times-Mirror Corp., is mostly 50-plus. Look at what the furniture manufacturers send their retailers: slicks, glossies and other graphics to be used in newspaper advertising. Hardly ever do they send out Radio commercials. If the furniture retailer buys a full-page ROP (run of paper) and pays megabucks for a full- or half-page ad, while newspaper's ADI circulation misses almost 60 percent of the market, and the retailer then has to discount its merchandise, how can retailers make any money?

To sum up the newspaper industry's attitude, Jim Wilson, VP of the Newspaper Advertising Bureau, was caught saying: "It's simply marketing 101. We're trying to get close to our customers again, something we didn't have to do when we were the only game in town."

The baby-boom generation, according to *Furniture Today*, the indus-

> - The baby-boom generation accounts for 52 percent of all furniture spending, ranging from first-time purchases to adapting to lifestyle changes.
> - As the end of the decade nears, aging boomers will shift away from high-end furniture to other priorities.
> - Radio is the dominant medium reaching baby boomers.
> - The furniture industry has been targeting the wrong population segment and using the wrong medium.

Chapter 44
The Furniture Boom

try's trade publication, accounts for 52 percent of all furniture spending. While the 25- to 34-year-olds contribute 22 percent of all consumer spending, they also account for 24 percent of all furniture expenditures. The 35- to 44-year-olds represent an even-greater buying clout, accounting for 28 percent of all furniture purchases. Does this ever-mobile baby boomer add up to 52 percent of all furniture purchases?

Then why are we doing such a poor job selling the furniture retailer when the 45- to 54-year-old customer, whom the furniture retailer has been targeting, represents a 10 percent drop in furniture spending? That age group contributed only 18 percent of all furniture spending, while the 55-64 age group dropped to 15 percent. Is their inability or unwillingness to change and move into the future the reason retail furniture dealers have to resort to almost giving their merchandise away by advertising: No payments for a year ... No down payment ... No interest for a year? If the industry has a bright future for the rest of the 20th century, can this be a way of increasing their margins?

Boomers' Buying Time

There are two underlying factors that account for why baby boomers, as a special group, buy so much furniture: First, there are 80 million of them. Second, the boomer is deeply immersed in household formation, such as buying homes, marrying and having children. These life stages will be important for the next five years, when the youngest boomers turn 35 and enter their peak earning years.

Additionally, baby boomers represent several short-term buying opportunities resulting from such activities as:

- The youngest boomers are in their initial furniture accumulation years — moving from rent-to-own to their first unfurnished apartments, condos or townhouses.
- Baby boomers are aging, and their children are moving from preschool to first grade, to elementary school, to high school and college. These different stages represent different furniture needs.
- Older baby boomers are making replacement purchases, including re-upholstering their family room or living room furniture.
- Young baby boomers will be marrying in droves in the next five to seven years.

The Key Decade

As older baby boomers enter the empty-nest period and as aging boomer parents find their children moving back home, the furniture needs of each different life-change will continue to impact the furniture industry's fortunes. Additionally, the swelling 35- to 44-year-old group of baby boomers will be buying new furniture to fit their newer, upgraded home purchases. Presently, there is much remodeling among the 30-something boomers who are moving into older homes. During the remaining years of this century, new marriages of DINKS (double income no kids) will represent 60 percent of new household growth, stimulating another time rich in furnishing purchases.

Another significant factor for the future of the furniture industry is that boomers will begin moving away from high-fashion furniture toward more functional, entertainment-oriented furnishing sparked by consumer electronic purchases.

As the end of the decade nears, aging boomers will shift away from high-end furniture toward providing educational needs for their children, health care for themselves and their aging parents, and a retirement nest egg for the 28 years longer they will live after retirement. Life-prolonging physical fitness research and breakthrough medical technology will cause boomers to seek ways to stretch their incomes to cover their longer life spans.

Right now, you can help your furniture dealer reach the growing masses of "baby boomers" while they are a captive audience in their vehicles. This mighty double-income consumer force is just beginning to form its buying preferences, and the newspaper industry has never marketed to them appropriately.

No other medium dominates the baby boomer as does Radio during the most critical buying periods of the day, 6 a.m. to 7 p.m.!

Stop talking ratings and why your Radio station is better than your competition. Instead, why not consider the impact of "integrative communications," which brings together various media voices to enhance the impact of the advertising effort? Why not recommend another Radio station as a complement to your station and the furniture dealer's current media selection? Why not recommend the furniture dealer use direct mail to its customers, which your station can promote? Why not recommend The Today Show, Good Morning America and several spots in late fringe news or spots on a Fox affiliate during the furniture merchant's sales spikes throughout the year?

Off Your Knees!

You say you are a "consultant." How are you improving Radio's image as a forceful medium when you tell the advertiser to use only Radio and to use only your station? Enough of the slam-dunk approach! Let's bring the full force of Radio and the burgeoning economic clout of baby boomers to the doorsteps of the merchant. As their business grows, so grows their budget and yours.

Chapter 44
The Furniture Boom

Stop selling two- or three-week flights. Long-term Radio campaigns create stability and credibility with the furniture dealer's present and potential customers.

Stop being subservient to the buyer. Get off your knees! As Radio's future unfolds, remember: Radio will still be the most omnipotent medium in reaching the target just before a buying decision goes down. If you help your furniture customers move into the future, wouldn't they want to take you with them into the future?

Chapter 45

Grill Your Salespeople
Before Your Clients Rake Them Over The Coals
By Chris Lytle

The No. 1 buyer complaint is lack of preparedness, according to a recent *Purchasing* magazine survey. So it would only make sense that sales managers who focus more training time on pre-call preparation than they do on closing will find their salespeople selling more.

Managers, Take Action

Try this little experiment. Ask each salesperson these questions today:
- What is your most important call today?
- What are you trying to accomplish?
- How far along are you in the sales cycle?
- What happened last time?
- Who are the key players?
- What information are you bringing the client?
- Why should the client do what you are suggesting?
- Have you consulted RAB information or done any outside research on their industry and customers?
- What is the biggest objection you anticipate, and how will you deal with it?
- What questions are you going to ask the client?

If your reps can't answer these questions smoothly, then the call they are about to make could be turbulent. Suggest to the rep that he or she

> - Lack of preparation is the No. 1 complaint buyers have about sellers.
> - Suggest postponing important calls rather than going unprepared.
> - Become passionate about preparation, and closing ratios will go up.
> - Control the call by controlling the information you bring to the client.
> - Fact-finding starts before you walk in the door.

Chapter 45
Grill Your Salespeople

not make the call. Request that they call the client and say: "I want to prepare better for the call so it is more productive for both of us." That will show them how serious you are about the quality of the call. By now, they already know how serious you are about the number of calls.

Pre-Briefing Is Preferred

This kind of coaching ("pre-briefing") is much more productive than debriefing a call. Either you grill your salespeople before they make the call, or the clients will rake them over the coals.

Sales success starts before you walk in the client's door. Building a solid, pre-call plan is crucial to your success.

Homework Opens Doors

Mack Hanan's advice: "If you don't have a plan, stay in the car," resonates. Opening the call with: "Is anything coming down for me this week?" is not a plan.

One salesperson told me that when he arrived at a scheduled appointment, the client was busy and said: "Just leave your package and I'll get back to you." This sales rep said: "My presentation is based on research I have been doing and requires some discussion."

"OK," said the client. They spent a productive half-hour together. Funny how telling the client you've done some homework can open doors and minds.

Salespeople who get in the habit of preparing have calls that are shorter and more productive. They find themselves getting down to business faster. And they set the standard for how Radio is sold in the market.

Information Is Power

The Wall Street Journal reports that *The Official Airline Guide* earned more money publishing airline schedules last year than the airlines made flying those schedules.

That goes to show how people are willing to pay for usable, targeted information. Information is power. Salespeople who prepare by researching information about the client's business gain tremendous power. They find that they can control the focus of the call and, thus, control the call.

Imagine the leverage. There are approximately 65,000 Radio reps in the United States. Let's assume they are making five calls a day. That's 1,625,000 calls a week. What an opportunity to prepare client-focused information and solid questions.

Media kits have never sold one second of broadcast time. And yet, stations strive to fill increasingly fancier folders with a stack of information that buyers ignore.

The key to success is a lot less glamorous: Do your homework. Finding facts that are relevant to your clients can position Radio reps as the best in the business.

Chapter 46

Emotional Strategy
The Other Half Of The Marketing Equation
By Ted Bolton

The day John Travolta stepped out on the disco dance floor will certainly go down in history. This one man, in one movie, created not only a new Radio format, but he virtually gave birth to a new way of life.

The same can be said of Al Ries and Jack Trout when they released their book, *Positioning: The Battle For Your Mind*. This one book has changed the face of Radio and created a culture of marketing warriors and a textbook of positioning credos.

It's easy to understand the Travolta/discomania phenomenon. Who could resist the pulsating disco beat and the white leisure suits? The Ries and Trout phenomenon is equally easy to understand.

> - Logic is only half the marketing story; the other half is emotion.
> - Emotional appeals can often bypass logic in the consumer's evaluation of your Radio station.
> - While most of the competition is still relying on the logic of warfare in marketing, now is the perfect opportunity to try the power of emotional marketing.

Airtight Logic

Ries and Trout succeeded in influencing the Radio industry by creating an airtight system of logic that is hard to argue with. Take, for example, their first rule of positioning: Just be first. It makes sense to believe that the first guy in has the leg up on the competition. How about the logic of developing and defending your position of strength or your "hill"? Equally logical and airtight. And, finally, how can you argue with "the perception is the reality"? It's the bylaw of marketing from which all strategies follow.

Chapter 46
Emotional Strategy

Simple? Yes. Does it give you the whole picture? No. In fact, it's exactly half of the marketing equation.

Getting To The Other Half

If you have ever experienced an emotion, then you already know that there is more to life than logic. The problem with positioning logic is that it completely ignores the way in which people are motivated and make decisions in a non-logical and emotional way.

The people who study how consumers make decisions can help us understand this process. First, let's begin with a typical logical decision. It could be represented by the accompanying diagram:

As the diagram shows, information is first evaluated, then it is either accepted or rejected. This evaluation process can be thought of as a buffer for reality. Information is either accepted or rejected, based on its truthfulness and its utility to the listener.

Where Ries and Trout leave us stranded is when this logical evaluation takes place. The problem is that after 10 or more years of warfare positioning strategies, the likes of "More Music," "The Biggest Variety" and "The World's Biggest Mix," the listener has begun to evaluate and reject these tactics as pure advertising hype. That means in the warfare logical system, the message is frequently lost, and no persuasion takes place. There just has to be a better way. Fortunately, there is.

Emotional Bypass

Consider what can happen with an emotional appeal. The quickest way to understand an emotional appeal is to think about how perfume is marketed. The manufacturers of perfume market love, sex and fantasy — pure emotion. Think about how a group of Radio broadcasters would market perfume. We would hear claims of "The Biggest Mix of Fragrances," or "Less Alcohol ... More Flowers." The sellers of perfume already know that emotion motivates and logic can only sometimes persuade.

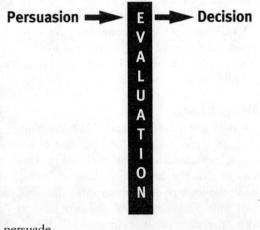

That means an emotional decision might look something like this:

Here you can see the power of emotional marketing. In this case, an emotional appeal can oftentimes completely bypass the evaluation buffers set up by a logical appeal. Examples of this abound in everyday life. Have you ever purchased a product because of an emotional impulse? Can you think of products that are sold on the basis of emotion? The fact of the matter is that Budweiser, Apple Computer, Pepsi, Chevy Trucks and Nike (to name a few) sell nothing but emotion. Why? Because they understand that logical positions get lost in the sea of competitive advertising claims.

The Opportunity Of A Lifetime

Radio is often slow to embrace change. That leaves most all of the competition over on the logical side of the marketing battlefield. Herein lies the opportunity of a lifetime to play in the uncontested arena of emotional marketing. The only obstacle is your creative imagination and your willingness to go where no person has gone before.

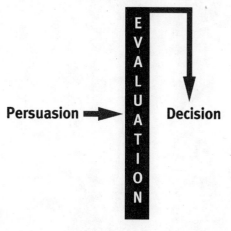

Think through the power of emotional marketing and tap into the other half of the brain. Use research to explore the emotional side of your Radio station and discover new ways to drive home listener loyalties that go far beyond "The World's Biggest Mix" or the "Best Variety." Look for emotional hot buttons that can be tied to your Radio station for a long-term effect.

If the way to a man's heart is through his stomach, then the way to a listener's mind is through the heart. And if you think that millions joined John Travolta out on the dance floor dressed in leisure suits because it made good, logical sense ... think again.

Chapter 47

Understanding And Selling Lifestyles
It Could Mean The Keys To The Kingdom
By Linda G. Brown

Just when you thought you'd never get direct information from Arbitron, the company increases sample size and plans to release lifestyle information on the ever-elusive diarykeeper. Understanding lifestyles and how to sell them is a new challenge for the Radio industry. For some, it may be the keys to the kingdom — if they know how to unlock its potential.

The Division Of A Market

Any discussion of lifestyles must begin with geography. Arbitron has been using zip codes for years to report Radio listening. In general, zip codes were convenient for Arbitron to deliver and easy for subscribers to understand. However, zip code boundaries are unrelated to the types of people enclosed. Zip codes are relatively unstable; they are created and destroyed at the discretion of the Postal Service. They do not include any set number of households, nor do they conform to the county boundaries of the Arbitron metro area. Although zip codes effectively divide the United States into optimum postal delivery units, they are a poor geography for most micromarketing techniques.

With rare enlightenment, the government designed a better system. The Office of Management and Budget (part of the Census Bureau) divides each county by tiers of smaller, more usable sample units. Counties are divided into census tracts, each representing approximate-

> - Arbitron has been using zip codes to report Radio listening, but a better system, devised by the government, divides counties' block groups, the smallest-acceptable geography for statistical analysis. This is the level at which Arbitron proposes to provide lifestyle data.
>
> - Lifestyle segmentation will provide stations with powerful information to use in targeting listeners and selling to advertisers.

Chapter 47
Understanding And Selling Lifestyles

ly 4,000 people or 1,500 households. Each tract is designed to represent meaningful socioeconomic areas for civil planning, which makes it a superior research geography.

The census tracts are divided by block groups of about 1,000 people (200 to 550 households). Block groups are further split into blocks, roughly equivalent to a city block (18 to 20 households) and bounded by streets, streams, railroad tracks, etc.

The 1990 Census was the first time that data was gathered for block groups and blocks across the entire United States. Block groups are considered to be the smallest acceptable geography for statistical analysis. This is the level at which Arbitron proposes to provide lifestyle data.

Lifestyle Segmentation Systems

Radio may be a mass medium, but stations and formats are not mass products. A station's image and programming attracts a select audience. Lifestyle segmentation gives definition to the Radio listener, providing numbers to support instincts.

With the aid of computers, vast amounts of data from the census and consumer surveys can be matched to geography, revealing distinct relationships between lifestyle and location. The process quantifies and qualifies the neighborhoods we already know — the "wrong side of the tracks," the "richy rich" and the "white-bread America." Industries that track consumers and their addresses use the systems to create customer profiles, which, in effect, define the customer base by stereotypes. These stereotypes tell what is likely to excite the consumer — powerful information that can determine product development, drive marketing strategies, expand trade areas and justify new site locations.

Several lifestyle segmentation systems are available today. Some of the most prominent are: PRIZM (Claritas/NPDC), MicroVision (Equifax), ClusterPLUS (Donnelley) and Lifestyles (National Demographics and Lifestyles). Also prominent in the industry are VALS (Marshall Marketing and Communications) and Stowell Data (Leigh Stowell and Company). Since the Arbitron Co. maintains a business relationship with Claritas/NPDC, it is possible that they will continue this alliance with PRIZM at the block group level. Although some may disagree, there is no lifestyle segmentation system that is best. All are based on legitimate research methods. All provide insight to the circumstance of the prospective consumer. The advantage to nationally defined neighborhood systems is that the system is consistent at any level of geography. On the other hand, a locally defined system is more likely than the national system to measure attitudes that may be unique to the market.

Another option is for Arbitron to design and market its own lifestyle system. No longer burdened by television ratings, Arbitron has more freedom to create a strong Radio-specific lifestyle system, but in today's competitive research environment there is some question about the merit of reinventing the wheel. The collection of demographic data on diarykeepers is feasible, but in order to compete with the other established systems, drastic methodology changes are needed. Already, there are undercurrents that Arbitron's ties with VNU, the parent company to both Claritas/NPDC and Scarborough, are weakening. Regardless of the final decision made under Arbitron's new progressive leadership, the demand on the Radio industry to understand and utilize lifestyle data will be the same.

How Will Arbitron Utilize Lifestyles?

Many questions surround Arbitron's release of lifestyle data: Will the new sample sizes statistically support the division of TALO into 40 (or 61) lifestyle groups? Will allowances be made for the difference in P1, P2 and P3 listeners? What flexibility will there be in receiving information? Will the lifestyle profile accurately represent a station's true following? How much will it cost?

It is unlikely that Arbitron will release the actual lifestyle or block group code specific to the diarykeeper, i.e., appended to a Fingerprint or Mechanical Diary. This type of information borders on invasion of privacy. Instead, as suggested by both Arbitron and Claritas representatives, it is likely that a station's "profile," a distribution of the 40 clusters, will be made available as a separate product.

Understanding the lifestyle distribution of diarykeepers will help you find ways to reach others like them. Since segmentation is linked to geography, direct mail can be targeted by specific lifestyles. Billboards can be placed in areas most likely to be noticed by your diarykeeper "types" in your hottest zip codes. Do you want your TV spot during the local news or the soaps? Your lifestyle profile can show ways to reach those who are predisposed to your programming. This stacks up to more-effective spending of limited marketing dollars.

Lifestyle information can also spark ideas on the theme and message. For, even if you reach the right people, you can't have an effect if your message doesn't impress them. Half of the job is finding your potential diarykeepers; the other half is moving them to action. Lifestyles can provide insight to both tasks.

Be Prepared

Arbitron is finally responding to the demands of the marketing environment with the introduction of lifestyles. Meanwhile, the Radio

Chapter 47
Understanding And Selling Lifestyles

industry is hailing its ability to now talk the advertiser's "language." There will be income distributions, education levels, occupations, purchasing habits and family status to describe your listeners — the same niche-selling tactics used by cable. Radio will have more than share and cost per point to bring to the table.

Be prepared to rethink current sales strategies. The wake-up call is this: Lifestyles will mean nothing unless your sales force can sell it. Otherwise, the "keys to the kingdom" may go to the competition.

Chapter 48

Copy That Sells
10 Tips For Making Radio Work
By Pam Lontos

If Radio doesn't work for a client, there are only two possible reasons — bad copy or improper schedule. The key to writing good copy is in knowing what information to get from the client, whether the salesperson writes the copy or gives that information to a copywriter.

Push The Right Buttons

When you sell a client, it is your responsibility to ask questions about their business so your commercials will push the client's hot buttons. If the client gives you copy you know is not

> - If Radio doesn't work for a client, it is because of bad copy or improper schedule.
> - The key to writing good copy is in getting the right information from the client.
> - Grab the listener's attention right away, and hit them with benefits.
> - Use hot words, such as money, improved, gain or love, for a better response.

going to work, it is your responsibility to inform them and help them rewrite it to get results. Second, it's your responsibility to run the client's ad when their business will get the best results.

Say to the client from the start: "Radio works! If this ad doesn't work for you, it is not Radio but your schedule or copy. Call me immediately so we can change the copy, schedule or both."

Also, let the client know Radio's strength is in its long-term, subliminal sell, that takes a few weeks to build. This way, they won't expect fast results like print, and cancel when they don't get them.

Clients will still want fast results, though. So how do you get them? Get away from "cutesy," overly creative ads that leave out benefits to the customer and contain no special offer to create immediacy.

Chapter 48
Copy That Sells

First, Grab Them

Here are 10 guidelines to follow when writing copy that brings immediate results.

1.) *The first sentence must grab the listener's attention.* For example: "Isn't it time you did something about your body?" or "Do you want more money?"

2.) *Hit them with benefits.* They don't care about computers, exercise equipment, a bank, etc. All they care about is: "What's in it for me?" Describe the product and they tune out. Tell them how it will help them and they turn up the volume.

Example:

Wrong: "The Nautilus cam gear has positive and negative resistance."

Correct: "Exercise on Nautilus and you'll firm up, lose inches, look and feel great!"

3.) *Take out non-meaningful words and fill with information that addresses the customers' needs.* Example: "If you are listening to this ad and you want new furniture …" The phrase "If you are listening to this ad" is unnecessary and takes time away from your 30 or 60 seconds, which should be used for selling.

4.) *Send them to White Pages instead of Yellow Pages.* If the ad says: "See our locations in the Yellow Pages," it sends the listener to the competition's advertising.

For A Limited Time

5.) *Offer a bargain and put a time limit on it.* People need a reason to act quickly. Often, customers won't even buy sale items but are drawn in by them and buy other products at full price. You must have a time limit or mention limited supply to create a sense of urgency.

6.) *Mention more than one item.* Every time you add a new item to an ad, you increase your chances of finding more customers.

7.) *Mention the phone number three times.* People usually don't have a pencil and paper with them when the ad airs.

8.) *Use the word "you" often.* People notice an ad with the word "you" because it directs the ad to them.

9.) *Use the word "because."* Customers react better when they know why something is marked down for a sale or why there is a limited supply.

10.) *Use hot words such as money, gain, improved, loss, love.* The reason so many products come out with "new, improved" versions is that this advertising technique works.

Recently, I was making calls with a salesperson. The client was a video store that had newspaper coupons for free video rentals for the weekend. He told me: "I'll test Radio by advertising early in the week when it's slow, and use print for the weekend."

I told him: "No. Reverse your strategy. Use your newspaper coupons early in the week and use Radio for your weekend. This will be a fairer test of Radio's pulling power compared to newspaper."

You must be in control of the client. You are in the Radio business and know what will work best for them. They want you to give them good advice to make their advertisement bring them higher returns.

Chapter 49

Furniture, Bridal And Bath
3 Markets To Sell To Boomers Now
By Dr. Philip J. LeNoble

The home furnishings industry will continue to grow as a revenue source for Radio, but certain segments of the industry — and certain age groups — hold particular promise.

The 44- to 55-year-old consumer, for instance, has other spending priorities, such as the care of aging parents, future retirement needs, the cost of health care and the cost of sending the kids to college. They will be less inclined to make expensive furniture purchases. So where will the future of upper-end home furnishings turn?

According to *Furniture Today* (Sept. 27, 1993), the high-end home furnishings retailer targets only 8.4 percent of the U.S. population. Only 4.4 percent of all households earn more than $100,000. That leaves a large chunk that will draw significantly from young singles and young marrieds in the range of $75,000 to $100,000. This segment has greater buying power now, before college funds and other concerns come into play.

Baby boomers who are between 30 and 48 years old present a real challenge for furniture retailers. They want more value and quality, but there is a trend away from luxury spending. If high-end retailers want to entice these consumers now, they must not pitch price alone but other benefits, such as what the furniture does for their career mobility.

- The best target market for high-end furniture is the segment of young singles or young married couples in the $75,000-$100,000 income range.
- Brides-to-be make up a significant market for home furnishings, and stations should encourage specialty stores to position themselves as bridal registries.
- Maturing baby boomers are expected to increase spending on bath-related products, opening up sales and promotion potential for retailers and manufacturers.

Chapter 49
Furniture, Bridal And Bath

The upper-end retailer is not concerned with price, since they are selling upper-price points. They just want to attract buyers to their store. Stations can capitalize on this by creating events that make the store a destination. Get away from cost-per-point and "we're No. 1." If your station's ratings are slim, but you're No. 1 in the cash register, then you really are No. 1.

Bridal Shower

Another potential market in the home furnishings industry is brides-to-be. In the coming years, "baby busters" (19-28 years old) and younger baby boomers (28-34) will be planning weddings and setting up households, so stations targeting these age groups should help specialty stores position themselves as the place to register for home furnishings.

According to *Home Textiles Today*, the better department stores constantly market themselves to brides; as a result, they own that front-of-mind awareness. The publication found that 59 percent of brides chose to register at better department stores like Macy's and Marshall Fields. Specialty stores fare worst (4 percent) because they do the least amount of marketing to the brides-to-be. As specialty stores grow in number, Radio stations should encourage them to position themselves as "The Registry." Doing so will allow them to gain great market share or else remain, as they have been for the most part, anonymous.

Bath Boom

The bath business is also a great new prospect in the baby boomer market.

According to *Home Textiles Today*, maturing baby boomers (35-44) are expected to spend $969 million for bath-related products in the next six years, a 37 percent increase in bath towels, rugs and shower curtains. That means it's time to start targeting linen and bath accessories retailers, department stores and manufacturers. Why not tie in a promotion with bedding and have a pajama party for young marrieds celebrating their 10th to 15th anniversary? Everyone who wears a pajama item gets to pick their favorite item and make an offer at the cash register. It can be an exciting way to create a mini-event that beats the traditional January White Sale. Best times are after the sun goes down on weekdays like Monday through Wednesday. Have a tie-in with a restaurant and cater the affair in exchange for advertising exposure, and have the local movie theater give away tickets to everyone who comes in.

Chapter 50

What Radio Taught TV
And How To Re-Learn It
By Cliff Berkowitz

Not that long ago, TV very smartly took its cues from Radio on how to promote its shows effectively. After all, TV was born out of Radio, so it seemed only natural that the most talented promotions people would come from Radio. Since then, somewhere along the way, Radio has forgotten most of the tricks of the trade, while TV has honed them to a science.

Do yourself a favor: Watch how TV stations promote themselves. You'll find they run promos every break. You'll also see them run promos back to back, in the middle of stop sets, all over the place. But aside from sheer volume of promos, there is also a method to their promo madness. You'll see them promote what's coming up in a half hour, later that night, later that week and something special down the road. This is good, effective promoting!

- Radio has forgotten most of the promotional tricks of the trade, while TV has honed them to a science.
- Always promote what's coming up later in the hour, later in the day, later in the week and something special coming soon.
- TV is best at teasing. Tease what's coming up in minutes to hold the listener's interest.
- Be more creative with promo scheduling. Run them in the middle of a stop set or two promos back to back.

We in Radio run promos once or twice per hour. We also tend to run about one or two liners an hour. And, generally, they are in equal rotation. Television has far fewer opportunities to promote itself than Radio, yet it does a phenomenal job at it.

T's, Tees, Tease!

What they do best is tease. By giving you just a taste of what's com-

Chapter 50
What Radio Taught TV

ing up, they keep you captive ... Coming up, a man who found something in his backyard that dates back to the Jurassic period. This also is promotion at its best. If you do the following, your time-spent-listening will increase:

• Always promote what's coming up later in the hour, later in the day, later in the week and something special coming soon.

• T's, Tees, Tease! Always be teasing what's coming up in "minutes." It doesn't have to be earth-shaking; it just has to sound like it. For example, don't say: "The new Janet Jackson song is coming up." Say,: "She was topless on the cover of a popular magazine last month, and her latest is coming up next!"

• Get "out of the box" when it comes to scheduling your promos. There is no law that says you have to run one or two promos an hour going into a stop set. Run more and shorter promos. Run some short teaser promos in the middle of stop sets. Try running two promos back to back occasionally.

• Log promos and liners. Don't just let them fall where they may. I've heard stations read a liner about something and segue into a promo on the same topic. Think! Get with your traffic manager to work out a system you can all live with.

Habit and complacency are the enemies of good Radio. Don't let industry standards dictate how to run your station. While you may be well-versed in what is "the right way" to do things, your listeners never got the memo.

If you are willing to experiment, perhaps you can teach TV a thing or two.

Chapter 51

From Ink To Air
10 Ways To Make Those Newspaper Dollars Yours
By Pam Lontos

Newspapers are taking away advertising dollars that should go to you. Radio gets just less than 7 percent of the advertising dollar; other media get the rest. It makes sense to go after newspaper advertisers for two reasons:

1.) Newspaper readership is on the decline. People today depend on electronic media.

2.) These non-readers are the "baby boomers" who comprise 44 percent of the 18-plus adults and account for almost 60 percent of consumer spending.

> * Fewer people read the newspaper, while more and more are listening to Radio. Yet Radio gets just less than 7 percent of the advertising dollar.
>
> * Sell all-newspaper advertisers by asking about their total advertising budget, not just what they spend on Radio.
>
> * Sell the long-term effects of Radio.

Selling Points

Here are some pointers on how to convert "newspaper dollars" to "Radio dollars." Remember, do not ever attack newspapers or someone's decision to use newspapers.

1.) *Call on accounts that aren't presently using Radio.* I once hired a lady who had no broadcast or sales background, and she got everyone else's rejects — the all-newspaper accounts. Within six months she was top biller, and she said that it was because I had given her such a good list. Everyone knows the all-newspaper accounts are the ones that spend the most money.

2.) *Ask about the client's total advertising budget.* If you ask a client what they spend on Radio, you may get an answer of several hundred dollars. However, if you ask for their total advertising budget, the answer will be several thousand dollars. Set your sights higher and start the conversa-

Chapter 51
From Ink To Air

tion on a higher monetary level.

3.) *Don't let budget objections get in the way of selling.* There's no such thing as a budget being spent; the excuse simple means that you haven't generated enough interest. And there's no such thing as a fixed budget. If you are talking to the decision-maker, he can change the budget at any time if he is convinced that it will add more profit.

Standing Out

4.) *Use psychology.* Concentrate on the fact that Radio will make their newspaper ad stand out. Talk about the Reticular Activating System at the base of our brain. It is a filter that cause us to block out some things and let other things through, such as when you hear your name among the 50 conversations at a crowded party. The same thing happens with newspaper ads. What causes us to notice certain ads? Those that we notice are the ones that we have been pre-programmed by Radio to notice. Tell the advertiser: "After you've been on the Radio for several weeks, you'll notice that your newspaper ad is working better." Now, when his customers tell him that they saw the ad in the newspaper, he'll give credit to his Radio message.

5.) *Demonstrate how Radio fits today's lifestyle.* In today's information age, people are bombarded with so much data that they don't have enough time to take it all in. It takes approximately 34 minutes to read a newspaper, and that requires pre-planning. Radio doesn't require this planning — people listen to Radio everywhere.

6.) *Sell the long-term effect of Radio.* Let clients know that they will get more response in the second month than the first, and the fourth month more than the second. Newspaper ads work over a period of approximately two days, so the all-newspaper advertiser expects this type of response from any advertising. Radio spreads its response over a period of several months. If you make sure the client understands this in the beginning, there will be fewer cancellations and more satisfied advertisers.

Subliminal Effect

7.) *Outline the types of customers reached by Radio.* Retailers are interested in three different types of customers: 1.) Those who need it and know they need it; 2.) Those who need it but don't know they need it and 3.) Those who don't need it now but will need it later. Newspaper appeals only to the first group. Radio's subliminal effect brings in all three groups.

8.) *Show the frequency comparison with newspaper.* How many times would that advertiser have to re-buy the newspaper ad in order to get the same frequency as Radio? The average is 50 Radio commercials for one full-page ad in the newspaper.

9.) *Sell yourself, then Radio, then your station.* A common mistake is to sell the station before the client is sold on Radio. Only when that client is sold on Radio are they interested in hearing about the specifics of your station.

10.) *Use facts and figures.* Gather all the subscription, readership and advertising information about the newspapers in your city and work out comparisons to your station's rate in advance. Calculate the population percentage who subscribe to the newspaper; you might be surprised at how low this figure is — and so will your clients. It's often 30 percent or less.

BIOGRAPHIES:

AMBROSE, ELLYN F. is CEO of The Marketing Group Inc. in Washington, D.C. She may be reached at 703-903-9500.

ANTHONY, DAVE is president of Anthony Media Concepts, a broadcast consulting company. He may be reached at 904-693-5235.

BACHMAN, KATY is a free-lance writer and editor. She may be reached at 203-353-8717.

BADER, MICHAEL H. is at the law firm of Haley, Bader & Potts. He may be reached at 703-841-0606.

BERG, MICHAEL is an attorney in Washington, D.C.

BERKOWITZ, CLIFF is president of Paradigm Radio, a Radio promotions and marketing consulting firm. He may be reached at 707-443-9842.

BOLTON, TED is president of Philadelphia-based Bolton Research Corp., a Radio research and marketing firm, and publisher of Radio Trends. He may be reached at 610-640-4400.

BOSLEY, RHODY is a partner with Research Director, Inc., a sales and marketing consultancy based in Baltimore. He may be reached at 410-377-5859.

BOYD, CLIFF is the president of Cash Flow Management. He has an extensive and diverse background in public and private financing and is passionately committed to providing the financial planning and partnerships necessary to allow all-size Radio stations to grow and prosper. He may be reached at 214-780-0081.

BROWN, LINDA G. may be reached through Eagle Marketing Services, Inc. at 303-484-4736.

BUNZEL, REED is vice president/communications at the Radio Advertising Bureau. He previously held the position of executive editor of Radio Ink, and has served in editorial capacities at Broadcasting magazine, Radio and Records and the National Association of Broadcasters. Bunzel is the author of "Pay For Play," a mystery novel published by Avon Books.

BIOGRAPHIES:

BURTON, BILL is president and COO of the Detroit Radio Advertising Group. He travels around the country touting his trademark and signature, "Be Fabulous," and is known for his crazy "stunts" and attention-getting ways. Burton holds a degree in business and economics from Michigan State University. He may be reached at 810-643-7455.

CARLOUGH, JUDY is the executive vice president of the Radio Advertising Bureau's national marketing efforts. She has more than 16 years experience in Radio sales and management at ABC RADIO, INFINITY BROADCASTING and NOBLE BROADCASTING, and for several years was an on-air personality in Boston. Prior to joining the RAB, she was the vice president and general manager of XTRA-AM. She may be reached at 212-387-2100.

CASE, DWIGHT is currently the president of Motivational Incentives Group and a member of WIMC's Executive Committee. He presently owns KOQO AM & FM and KSQR AM & FM, and has a 35-year history in broadcasting. His journalism experience includes four years as the publisher and chief executive of Radio & Records. He may be reached at 310-854-7505.

CHAPIN, RICHARD W. holds the distinction of being the only person ever to be elected as chairman of both the Radio Advertising Bureau and the National Association of Broadcasters. He has been recognized as R.C. Crisler & Company's leading volume broker, and has received numerous broadcasting awards and citations throughout the years.

CIMBERG, ALAN is a noted sales motivational speaker and trainer. He may be reached at 516-593-7099.

COLOMBO, GEORGE is a nationally recognized speaker and writer on sales and management topics. He is a regular columnist for Selling magazine and the author of the McGraw-Hill best seller "Sales Force Automation." Colombo has shared his message with thousands of sales professionals in workshops and presentations nationwide. He may be reached at 407-327-2453.

COOKE, HOLLAND is a Washington, D.C.-based programming consultant specializing in news/talk and full-service AM. He may be reached at 202-333-8442.

CORNILS, WAYNE is a vice president at the Radio Advertising

Bureau. He may be reached at 212-254-2142.

CRAIN, DR. SHARON is a leading industrial psychologist, educator and author. Her client list includes National Association of Broadcasters, USA Today, American Association of Advertising Agencies and Radio Advertising Bureau. Crain has appeared on The Today Show, Good Morning America and many others. She may be reached at 602-483-2546.

DONALDSON, MIMI is an experienced management consultant, trainer and speaker. She spent 10 years as a staff trainer in Human Resources at Walt Disney Productions, Northrop Aircraft and Rockwell International, and has been president of her own consulting company for the past 10 years. She may be reached at 310-273-2633.

FELLOWS, JOHN, CRMC, Mr. Radio™, is highly regarded as an author, speaker and sales professional. His practical sales and advertiser workshops have proven beneficial for associations, groups and stations. His latest book and workshop, "How To Get Rewarding Results With Radio Advertising," is now available. He may be reached at 800-587-5756.

FISHER, GARY is currently vice president and general manager at SW Radio Networks. Previously, he was the vice president and general manager at WNIC/WMTG Detroit and WHTZ New York, and the vice president and general sales manager at WABC MusicRadio/TalkRadio New York. He may be reached at 212-445-5409.

FRIEDMAN, NANCY, the Telephone "Doctor"®, is recognized as "America's formost and most sought-after speaker on customer service and telephone skills." She has spoken before many of the most prestigious associations and organizations around the world, and has appeared on many leading Radio and television programs. She also has a best-selling video and audio training library currently available in eight languages. She may be reached at 800-882-9911.

GABLE, CHRIS of Chris Gable Broadcast Services, a national Radio consulting firm, offers support for programming, management, marketing and development. He may be reached at 717-964-3255.

GALLAGHER, ANN may be reached at 202-619-2189.

GALLAGHER, GINA may be reached at 414-272-6119.

BIOGRAPHIES:

GIFFORD, DAVE is a sales and management consultant from Santa Fe, New Mexico. He may be reached at 1-800-TALK-GIF.

GOULD, MARTY may be reached at 419-228-9248.

HERWEG, ASHLEY and GODFREY are international seminar leaders who have owned, operated and managed stations in small, medium and large markets. They have also co-authored the informative "MAKING MORE MONEY ... Selling Radio Advertising Without Numbers" and "Recruiting, Interviewing, Hiring and Developing Superior Salespeople." Both may be reached at 803-559-9603.

HESSER, MICHAEL B. may be reached at 805-543-9214.

HOFBERG, BUNNY may be reached at 212-613-3816.

KARL, E. is president of E. Karl Broadcast Consulting, a Radio programming and marketing firm. He may be reached at 805-927-1010.

KEITH, MICHAEL C. is a member of the communications department at Boston College. He has also held various positions at several Radio stations and served as the Chair of Education for the Museum of Broadcast Communications. Additionally, he is the author of several books on electronic media, including "Signals On The Air," "The Radio Station" and The Broadcast Century." He may be reached at 617-552-8837.

KNOX, BRIAN K. may be reached through Interep at 212-818-8933.

LeNOBLE, DR. PHILIP J. is chairman of Executive Decision Systems Inc., a human resource, sales training and personal development company in Littleton, Colorado. He may be reached at 303-795-9090.

LONTOS, PAM, president of Lontos Sales & Motivation Inc., customizes seminars, keynotes and "in-station" consulting for stations or associations. She may be reached at 714-831-8861.

LUND, JOHN is president of Lund Media Research and The Lund Consultants to Broadcast Management, Inc., a full-service Radio research, programming and consulting firm dedicated to assisting Radio stations achieve better programming, higher ratings, greater revenue and increased profitability. He may be reached at 415-692-7777.

LYTLE, CHRIS, president of The AdVisory Board Inc., is author of the Radio Marketing Master Diploma Course. He may be reached at 800-255-9853.

MAGUIRE, KATHRYN is president of Revenue Development Systems. She may be reached at 617-589-0695.

MAKI, VAL serves as vice president/general sales manager for WKQX-FM Chicago and has been involved with sales development for Chicago's EMMIS Broadcasting for the past 10 years. She may be reached at 312-527-8348.

MARTIN, ANDREA may be reached at 206-443-9400.

McDANIEL, MIKE, producer of the Action Auction promotion nationwide, has written a book about promotions, and owns and operates two Radio stations. He may be faxed at 812-847-0167.

OTT, RICK is president of the management consulting firm Ott & Associates in Richmond, Virginia, and author of "Unleashing Productivity!" and "Creating Demand." He may be reached at 804-276-7202.

PRESSMAN, ROY is director of engineering for WLVE/WINZ/WZTA in Miami. He may be reached at 305-654-9494.

RATTIGAN, JACK M., CRMC, is president of Rattigan Radio Services, a management and sales consulting company headquartered in Portsmouth, Virginia. He conducts his "The Basics & Beyond" one-day seminars in various markets. Previously, he worked in Philadelphia for NBC, Group W and Metro Media. He may be reached at 804-484-3017.

SABO, WALTER is president of Sabo Media, a management consulting firm based in New York, specializing in turnaround strategies for major market stations. He may be reached at 212-808-3005.

SISLEN, CHARLES may be reached at 212-424-6417.

SKIDELSKY, BARRY is an attorney and consultant, concentrating his efforts in the Radio industry. A frequent author and speaker, he is licensed to practice law in New York and Washington. His background includes 15 years in Radio programming, sales and management. He may be reached at 212-832-4800.

BIOGRAPHIES:

SUFFA, WILLIAM P., P.E. is vice president and management principal of Suffa & Cavell, Inc., with 18 years of experience in Radio communications systems engineering to his credit. As a private consultant, he has provided services for more than 450 clients in the telecommunications and broadcasting fields. Suffa is an author of a monthly column in Radio Ink. He may be reached at 703-591-0110.

TROUT, JACK is president of Trout & Ries marketing strategists in Greenwich, Connecticut. He may be reached at 203-622-4312.

WHITAKER, GEORGE is the author and publisher of Practical Radio Communications, a monthly newsletter that teaches Radio engineering to beginners. He has served as chief engineer for KRVA-AM/FM, Dallas, and is author of the book "Radio Engineering for the Non-Engineer: What Managers Need to Know About Engineering." He may be reached at 800-572-8394.

ZAPOLEON, GUY is president of Zapoleon Media Strategies and works with associates Jeff Scott and Steve Wyrostok. He may be reached at 713-980-3665.

SALES AND MARKETING INDEX

A

Account List Management System 16
Account
    assignments 51-52
    attrition 15-16, 39
    draft 55-57
    management 59
    new 15-17
    profile form 59-61
    retention 71-73
ACT! for Windows, DOS and HP95LX 135
ADI 163
Advisory Board 16
Age Wave 163
Aikman, Troy 64
Almond, Alan 148
American Banker 48
American Demographics 157
AmiPro 136
Anheuser-Busch 7
Apple Computers 173
Appointments (getting them) 111-113
Arbitrend 65, 147
Arbitron 23-25, 124, 176-177
    diaries 25, 175, 177
Assertive communication 119-121
AT&T 21
Automotive 1, 11, 33, 39, 157-158

B

Baby boomers 163-166, 183-184, 187
Baby busters 184
Bath 183-184
Berkowitz, Cliff 185-186
Beverage News 48
Billboards 177
Billing 15-17, 19, 38-39, 55, 71-73
Blore, Chuck 2
Boehme, Gerry 142
Bolton, Ted 23-25, 63-65, 171-173
Bonus spots 75

SALES AND MARKETING INDEX

Bookkeeper..15, 72
Bottom Line Communicator, The..................................39
Bouvard, Pierre...142
Brandweek...157
Bridal..183-184
Brown, Linda G..175-178
Browner, Michael...142
Budget cuts...153-156
Budweiser..173
Burns, George...125
Burton, Bill..1-3, 141-142
Buyers
   agency...41, 48, 60, 111
   local-direct...41
   media..90

C

Cable...46, 48
Cable Ad Bureau..46
Cadillac...47
Campbell-Mithun-Esty...141
Career stability..159-160
Carlough, Judy...153-156
Carson Pirie Scott...39
Case, Dwight..75, 76
Census Bureau..175
   block groups...176
   blocks...176
   tracts..175-176
Chapin, Richard W..19-21
Chevrolet "Heartbeat" campaign....................................2
Chevy..47
Chevy trucks..173
Churn...16, 71-73
Churn Calculation Worksheet................................72-73
Cimberg, Alan..31-35
Claritas/NPDC...176
Closing..123-124
Clothing...11, 157-158
ClusterPLUS..176
Coca-Cola...93
Cold calls..112

Collections 20, 46, 77-78
Colorado Rockies 55
Commence 135
Commissions 12, 16, 19-20, 45-47, 81, 103-105, 155
Commitment 63-65
Communication 51-52, 119-121
Community involvement 9, 39
Competitors
    battling them 143-146
    know their strengths and weaknesses 9
Computer Reseller News 157
Computers 135, 137, 158
Conlon, Jim 2
Consultants 81, 165
Contact management software 135-136
Convenience stores 60
Conventions 48
Co-op advertising 27-29, 155
    dollars 11-12
Copy book 155
Copywriting 42, 153-156, 179-181
Core Competencies 5-9
Corel Draw 136
Cost per point (CPP) 90, 93, 141-142
Counseling 51-53
Courtesy 139-140
Crain, Dr. Sharon 107-109, 127-129, 159-160
Cravens, John 141
Creativity 1-3, 41-43
Critical Success Factors (CSF) 159-160
Cume 141
Customer Needs Analysis 94
Customer retention 37-40
Customer service 37-40, 45-46, 97-99,131-133, 159-160

D

Dallas Cowboys 63-65
Darwin, Charles 55
Dayton's Department Store 26
Demo pricing 141-142
Demo tapes 2
Detroit Media Directors Council 141

SALES AND MARKETING INDEX

Department stores 184
Detroit Radio Advertsing Group (DRAG) 2, 141
DINKS (double income no kids) 165
Direct mail 48
Directory of Radio Computer Software Suppliers 137
Distributors 26
Do-it-yourself 157
Donaldson, Mimi 97-99, 119-121
Donnelley 176
Drug Store News 157
Duopolies 149
Dychtwald, Dr. Ken 163

E

"Earshot" 23-24
Eastman Radio 142
EDLP (every day low pricing) 158
Effective frequency 142
Emotional marketing 171-173
Empathy 97, 99
Environmental concerns 157-158
Equifax 176
ESPRIT 24

F

FDA 158
Federal Express 46
Fellows, John 37-40, 67-69, 101-102, 135-137
Fingerprint 177
First impressions 127-129
Fisher, Gary 147-151
Florida Marlins 55
Follow-up service 29, 45-46, 131-133
Food business 11, 157-158
Ford 86
Formats 12, 148
Fox network 165
Framemaker 136
Fraudulent billing 25-27
Fredrick, Ron 141
Free-lancers 155
Frequency comparisons 188

Friedman, Nancy 131-133
Friendship-based selling 20
Furniture 163-166, 180, 183-184
Furniture Today 163-164, 183

G

Gallagher, Gina 81-83
General Motors 93, 142
"Get Me Radio" 2
Gifford, Dave 77-79
Good Morning America 165
Gourmet magazine 47
Graphics programs 136
Grid card 47, 75
Grooms, Billy 41
Group W Radio 3

H

Hallmark Cards 93
Hanan, Mack 168
Handling rejection 101-102
Harper and Gannon 148
Harvard Business School 2
Harvard Graphics 136
Harvard University 5
Helene Curtis 60
Herweg, Godfrey and Ashley 93-95
Hesser, Michael B. 85-86
Hiring 19-20, 63-65
Holiday Inn 42
Home Depot 158
Home improvement 11, 157-158
Home Textiles Today 184
Hopkins, Tom 68, 102
How To Master The Art of Selling 102
Hypertarget 147

I

I Ching, the Book of Changes 98
IBM 21, 93
Industrial Revolution 93
Interns 82

SALES AND MARKETING INDEX

Investment..63-65

J

J. Walter Thompson 141-142
Johnson, Jimmy..63-64
Jones and Healy..43
Jones, Jerry..63-65

K

Katz Radio Group...142
KFS (key factors of success) 5-9

L

Landry, Tom..64
Lant, Dr. Jeffrey...125
Learning International...161
Leigh Stowell and Company..176
LeNoble, Dr. Philip J.................41-43, 87-88, 163-166, 183-184
Liddy, G. Gordon...24-25
Life's Little Instruction Book..123
Lifestyles...176
Lifestyle segmentation..175-178
Lintas: Campbell-Ewald 1-2, 141
LMAs...41
Lontos, Pam..111-113, 179-181, 187-189
Lotus Agenda..135
Lotus Freelance...136
Ludwig, Bill...1-2
Lytle, Chris...15-17, 55-57, 71-73, 89-91, 123-125, 161-162, 167-169

M

Macy's..184
Maguire, Kathryn L. 11-13, 115-117, 157-158
Maki, Val..27-29, 59-61, 103-105
Managers
    brand..60
    district sales..60
    local sales 12
    marketing..60
    merchandise 60
    regional sales 60
Mancini, Bob...142

Manufacturers/manufacturer accounts 11-13, 27-28, 59-61,
                                          104, 115-117, 157-158, 184
Marshall Fields 39, 184
Marshall Marketing and Communications 176
Marx, Steve 142
Mass marketing 93-95
Match (between salesperson and prospect) 51-52
McDonald's 39
Mechanical Diary 177
Media kits 124, 168
Mercury Awards 3, 155
Miami Hurricanes 63
Michelangelo 154
Michelin 158
Microsoft Word 136
Microsoft Powerpoint 136
MicroVision 176
Minnesota Vikings 64
Miracle Sales Guide 43
Mitchell, Bob 141
Motivation 51, 53
Mystery Shopper 115-116

N

National Demographics and Lifestyles 176
Nautilus 180
Negotiating 31, 34-35
New business development 81-83
New York Times, The 63
Newspaper advertising 86, 124, 154, 163-164, 181
    circulation 93
    cost vs. Radio 16
    revenues vs. Radio 46, 187-189
    tear sheets 85-86
Newspaper Advertising Bureau (NAB) 45, 137, 155, 163
NFL 63-65
Niche marketing 47, 93-95, 147-149
Nike 173
Nutrition 157-158

O

Odiorne, George 125

SALES AND MARKETING INDEX

Office of Management and Budget 175
Official Airline Guide, The 168
One-minute manager 150
Optimism vs. pessimism 51-52
Optimum Effective Scheduling (OES) 56, 142
Orkin, Dick 2
Overnights 23-24

P

"Passive Measurement System" (PMS) 23-25
Pepsi 86, 173
Perceived value 32-34
Persistence 67-69
Personal Information Managers 135-136
Peters, Tom 140
Phar-Mor 26
Pike, Robert 161-162
Pillowtalk 148
"Pocket People Meter" (PPM) 23-25
"Pocket People Person" (PPP) 23-25
Polo 24
Pool stations 154-155
Popiel Pocket Fisherman 23
Positioning 171
Positioning: The Battle For Your Mind 171
Postal Service 175
PR departments 60
Pre-call objective sheets 15-17
Pre-call preparation 167-169
Pre-empted commercials 75
Premium items 48
Price
    how to handle objections 31-35
PRIZM 176
Procter & Gamble 49, 59, 158
Profit 73
Programming 41, 46, 48-49, 149
Promotions 115-117, 185-186
    department 105, 149
Proof of performance 27-29
Purchasing 167

VIII   **Sales & Marketing**

R

Radio Advertising Bureau (RAB) 3, 45, 49, 87, 155, 167
   Instant background 162
"Radio Works" 2
Ragan, Brian 39
Rapport building 107-109
Rates 161-162
   reduced 20, 42, 75
   reasons to raise them 45, 47, 49
Ratings 23-25, 41-42, 46
   programs 137
Rattigan, Jack M. 139-140
Reach & frequency reports 137
Recycling 158
Reichenheld, Frederick 71
Rep firms 26
Restaurants 162
Retailers/retail accounts 27-28, 47, 59-61, 75, 104,
   .... 111, 115-117, 149, 157-158, 183-184, 188
Reticular Activating System 188
Ries, Al 171
Risks 63-65
RKO Radio 47
Robbins, Tony 68
Rogers, David J. 5-9, 51-53
Rolex 32

S

Sabo, Walter 45-49
Sales principles
   Defensive 143-144
   Flanking 143, 145
   Guerrilla 143, 145-146
   Offensive 143-145
Scarborough 177
Schedule 179-181
Schroeder, Randy 141
Schwartz, Tony 2, 123
Selling
   performance requirements 20
   rules 20

SALES AND MARKETING INDEX

Seminars 31, 42, 48, 81, 87, 101
Service companies 104
Service Edge, Minneapolis 131
7-Eleven 7
Signage 115-116
Sislen, Charlie 142
Sizzle letters 115-116
Sizzle tapes 115-116
Standard Rate and Data Spot Radio Guide 88
Stew Leonard's Super Market, Norwalk, Connecticut 140
Stowell Data 176
Strata 41-42
Strategic planning 5
Super Bowl 63
Supermarket News 157
Swatch 24
SWOTs (strengths, weaknesses, opportunities and threats) 5-9

T

TALO 177
Tapscan 41-42
Tear sheets 85-86
Teases 185-186
TeleMagic 135
Telemarketing 48, 135-136
Telephone 111
Texas Stadium 64
"Theater Of The Mind" 2
Thompson, Jim 3
Three-step method for saying "no" 119-121
Times-Mirror Corp. 163
Today Show, The 165
Total Quality Management (TQM) 159-160
Trade publications 40, 48, 157-158
Trade shows 48
Tracy, Brian 68
Training 19-21, 51-53, 78, 87-88, 155, 161-162
Transactional business 81
Travolta, John 171, 173
Trout, Jack 143-146, 171
Turnaround 147-151
Turnover 1, 20-21, 71-73

TV 48, 93, 177, 185-186
Typologies 51-53

U

"Unaided recall" 23-24
U.S. Department of Labor 45
U.S. Steel 93
USA Today 157

V

VALS 176
Value-added 76
Vendor sales 27, 59-61, 103-105, 149
   director 11-12
   dollars 11-13
Veronis-Suhler Media Report 45
Video stores 181
Vision 63-65
Visualization 1, 42
VNU 177
Von Oech, Roger 43

W

Walker, Herschel 64
Walkman 23-24
Wall Street Journal, The 24, 131, 157, 168
West, Bill 2
Whack On The Side Of The Head, A 43
White Pages 180
WHLP, Baltimore 94
WHYT, Detroit 141
Wholesalers 27-28
Wilson, Jim 163
WMFX, Columbia, South Carolina 41
WNIC, Detroit 148
Women consumers 157-158
Women in Radio 127-129, 159-160
Women's Wear Daily 157
Word Perfect 136
Word processing programs 136
World Series 55

SALES AND MARKETING INDEX

Y
Yellow Pages 149, 180

Z
Z100 21
Ziglar, Zig 67-68

RADIO INK BACK ISSUES AVAILABLE!

The issues of *Radio Ink* that you've missed are now available in limited supply. Hundreds of moneymaking ideas, interviews, sales tips, copy ideas, packages, marketing strategies and more that you can use now!

VISA, MASTERCARD and AMERICAN EXPRESS accepted!

Normally $4.50 each!
1 to 3 issues $4 each ($2.50 S&H)
4 to 6 issues $3.50 each ($3.50 S&H)
7 to 10 issues $3 each ($4.50 S&H)
11 or more $2.50 each ($6.50 S&H)
(Florida residents add 6% sales tax.)

#1 Jan. 8, '90
Cover: The Future Of Radio - A Look At The Decade Ahead
Interview: Dick Harris

#2 Jan. 15, '90
Cover: What The RAB Can Do For You:
Interview: Robert Sillerman

#3 Jan. 22, '90
Cover: Radio Commercials On TV
Interview: Carl Wagner

#4 Jan. 29, '90
Cover: An Arbitron Radio Diary
Interview: Jerry Cliffton

#5 Feb. 5, '90
Cover: Radio Group Heads (What They Look For When Hiring A GM)
Interview: Frank Wood

#6 Feb. 12, '90
Cover: Taking Over As GM
Interview: Ken Swetz

#7 Feb. 19, '90
Cover: Strange Bedfellows? (When Radio Owners Own Another Business)
Interview: Bob Fuller

#8 Feb. 26, '90
Cover: Pork Rinds And Porsches (Country Radio Goes To Town)
Interview: Bob Meyer

#9 Mar. 5, '90
Cover: Doing Remotes Fron Fantasyland (What Disney Has To Offer)
Interview: Steve Berger

#10 Mar. 12, '90
Cover: We Can Help (Executive Search Firms Answer Radio's Questions)
Interview: Aaron Daniels

#11 Mar. 26, '90
Cover: Hiring Sales Superstars
Interview: Carl E. Hirsch

#12 Apr. 9, '90
Cover: Employer Expectations And Employee Rights (How Much Should Be In Writing)
Interview: Marc Guild

#13 Apr. 16, '90
Cover: Power Collections
Interview: Ted and Todd Hepburn

#14 Apr. 23, '90
Cover: Training Radio Superstar Salespeople
Interview: Art Carlson

#15 Apr. 30, '90
Cover: Could Cable Sales Hurt Radio?
Interview: Tom Gammon

#16 May 14, '90
Cover: Researching The Researchers
Interview: Scott Ginsburg

#17 May 21, '90
Cover: Buying Your First Station: A Primer
Interview: Alan Box

#18 Jun. 4, '90
Cover: Rise Reported in Listening Levels
Interview: Al Sikes

#19 Jun. 11, '90
Cover: Hiring A PD
Interview: Jerry Lyman

#20 Jun. 18, '90
Cover: Back To School
Interview: Robert Kipperman

#21 Jul. 2, '90
Cover: EZ Listening: Eye of the Storm
Interview: Herb McCord

#22 Jul. 9, '90
Cover: A Look At The Direct Mail, Telemarketing for Radio
Interview: Raul Alarcon

#23 Jul. 16, '90
Cover: National Business
Interview: Mike Oatman

#24 Jul. 23, '90
Cover: Handcuffed By Your Sales Image
Interview: Michael J. Faherty

#25 Jul. 30, '90
Cover: New Life For News/Talk
Interview: Randy Michaels

#26 Aug. 6, '90
Cover: Traffic And Billing Systems
Interview: Jim Duncan

#27 Aug. 13, '90
Cover: Marketing Your Radio Station (How to Formulate A Strategic Plan)
Interview: Les Goldberg

#28 Aug. 20, '90
Cover: Effective Budgeting
Interview: Ted Nixon

Sales & Marketing

TO ORDER BACK ISSUES: CALL 1-800-226-7857

#29 Aug. 27, '90
Cover: Twenty-Four Hour Syndicated Programming
Interview: Jay Cook

#30 Sep. 3, '90
Cover: New Technology For Radio
Interview: Alexander Williams

#31 Sep. 17, '90
Cover: Goodbye To Boston
Interview: Jim Thompson

#32 Sep. 24, '90
Cover: '90 Marconi Awards
Interview: Bob Hughes

#33 Oct. 1, '90
Cover: Helping Your Clients Develop Marketing Strategies
Interview: Steve Edwards

#34 Oct. 8, '90
Cover: Sales Presentations
Interview: Frank Osborn

#35 Oct. 15, '90
Cover: A Niche In Time
Interview: Dick Ferguson

#36 Oct. 29, '90
Cover: When To Walk
Interview: Nick Verbitsky

#37 Nov. 5, '90
Cover: Correction or Catastrophe: (The Year In Trading)
Interview: Jeffrey E. Trumper

#38 Nov. 12, '90
Cover: Sales and Management Consultants
Interview: Stanley Mak

#39 Nov. 19, '90
Cover: Black-Owned Radio
Interview: Pierre Sutton

#40 Nov. 26, '90
Cover: Back to Basics and Beyond
Interview: Mickey Franko

#41 Dec. 3, '90
Cover: Libraries and Custom Commercials
Interview: Michael Bader

#42 Dec. 10, '90
Cover: Managing A Radio Station In A Recession
Interview: Dan Mason

#43 Jan. 14, '91
Interview: Ralph Guild (Radio Executive of the Year)

#44 Jan. 21, '91
Cover: Making TV Work Harder For Radio
Interview: Rick Buckley

#45 Feb. 4, '91
Cover: Tools That Make Your Station Sound Great
Interview: Rick Dees

#46 Mar. 18, '91
Cover: Hot Sales Prospects
Interview: Mark Hubbard

#47 Apr. 1, '91
Cover: Interactive Phone Systems
Interview: Bill Steding

#48 Apr. 29, '91
Cover: Advice From Great Operators
Interview: Marty Greenberg

#49 May 13, '91
Cover: Selling Car Dealers
Interview: Sally Jessy Raphael

#50 May 27, '91
Cover: Research Strategies
Interview: Terry Jacobs

#51 Jun. 10, '91
Cover: Changing Face Of Radio Engineering
Interview: Dick Kalt

#52 Jul. 15, '91
Cover: What Do Media Buyers Think Of Radio Salespeople?
Interview: Richard Balsbaugh

#53 Jul. 29, '91
Cover: Pilgrimage To Arbitron
Interview: Frank Scott

#54 Aug. 12, '91
Cover: Where Does Your Motivation Come From?
Interview: Rush Limbaugh

#55 Sep. 23, '91
Cover: Building A Competitive Advantage
Interview: Jacqui Rossinsky

#56 Oct. 7, '91
Cover: Direct Marketing For Radio
Interview: Warren Potash

#57 Oct. 21, '91
Cover: DAB: How Will it Affect Us?
Interview: Joe Field

#58 Nov. 4, '91
Cover: 24-Hour Formats
Interview: David Rogers

#59 Nov. 18, '91
Cover: How To Solve Your Biggest Sales Problem
Interview: Robert F. Callahan

#60 Dec. 2, '91
Cover: Sales & Management Consultants
Interview: Carl C. Brazell Jr.

#61 Dec. 16, '91
Cover: The Year In Review
Interview: '91 Interview Review

#62 Jan. 6, '92
Cover: The History Of Group W Radio
Interview: Jim Thompson

#63 Jan. 20, '92
Cover: How To Write Great Radio Spots
Interview: James H. Quello

#64 Feb. 3, '92
Cover: Breaking The 6.8 Barrier
Interview: Gary Fries

#65 Feb. 17, '92
Cover: How To Sell Retailers
Interview: Bill Livek & Bill Engel

#66 Mar. 2, '92
Cover: Country Radio
Interview: Jerry Lee

#67 Mar. 16, '92
Cover: Hit Promotional Items
Interview: Steve Marx
Interview: Pierre Bouvard

#68 Mar. 30, '92
Cover: New Technology
Interview: Neil S. Robinson

#69 Apr. 13, '92
Cover: LMAs
Interview: Barry Umansky

#70 Apr. 27, '92
Cover: Collection Strategies
Interview: John Dille

#71 May 11, '92
Cover: Computerization Of Radio
Interview: Gary Stevens

#72 Jun. 8, '92
Cover: NAB Radio Montreux
Interview: Dick Clark
Interview: Nick Verbitsky

#74 Jun. 22, '92
Cover: Choosing Programming Consultants
Interview: Gordon Hastings

(Back issues prior to Aug. 10, '92 are Pulse of Radio issues.)

Sales & Marketing   xv

TO ORDER BACK ISSUES: CALL 1-800-226-7857

Offer based on availability.)

#75 Aug. 10, '92
Cover: AM Survival Strategies
Interview: David Kantor

#76 Nov. 2, '92
Cover: Radio Revenues
Interview: Jimmy de Castro

#77 Dec. 14, '92
Cover: The Year In Review
Interview: The Best Of '92 Interviews

#78 Jan. 4, '93
Cover: Short-Form Programming
Interview: Gary Fries

#79 Mar. 1, '93
Cover: Country Radio
Interview: Ken Greenwood

#80 Mar. 29, '93
Cover: Sports Radio
Interview: Paul Fiddick

#81 Jun. 7, '93
Cover: The Future of Formats
Interview: Bob Sillerman

#82 Jul. 12, '93
Cover: Making The Move From PD To GM
Interview: G. Gordon Liddy

#83 Aug. 9, '93
Cover: Increase Sales With Software
Interview: Bob Fox

#84 Aug. 23, '93
Cover: Traffic & Billing Automation
Interview: Dan Mason

#85 Sep. 6, '93
Cover: Great Copy on a Limited Budget
Interview: George Carlin

#86 Oct. 4, '93
Cover: The Marketing of Urban/Black Radio
Interview: Steve Morris

#87 Nov. 1, '93
Cover: Selling Cost Per Point
Interview: Gordon Hastings

#88 Nov. 15, '93
Cover: New Technology Review

Interview: Wayne Vriesman

#89 January 3, '94
Interview: Mel Karmazin

#90 Jan. 17, '94
Cover: Talk Radio
Interview: Hank Stram & Jack Buck

#91 Jan. 31, '94
Cover: Spanish Language Radio
Interview: Cary Simpson

#92 Feb. 14, '94
Cover: The Ultimate Sales Manager
Interview: Ralph Guild

#93 Feb. 28, '94
Cover: Marketing Country Radio
Interview: Frances Preston

HOW TO ORDER:

Within the U.S. and Canada, call 1-800-226-7857 or 407-655-8778 with your credit card information between 9 a.m. and 5 p.m. Eastern time.
Outside the U.S., call 407-655-8778.

To order by FAX:
Fax form with your credit card information and signature to 407-655-6164.

To Order by Mail:
Mail this form with your check or credit card information to:
Streamline Publishing, Inc., 224 Datura Street, Suite 718
West Palm Beach, Florida 33401-9601, U.S.A.

MAIL OR FAX ORDER FORM:

☐ Please enter my subscription to *Radio Ink* magazine, published every other week (25 times a year). The cost is $125.00 ($199 International.)

☐ Please send me _____ copies of *THE RADIO BOOK*™: *The Complete Station Operations Manual* (a three-book set) for $89.00 for all three (plus S&H and tax where required). I understand that if I am not satisfied with any book for any reason, I may return it within 30 days.

$89.95 each set $_____

Fla residents add sales tax (6%) $_____

Shipping & Handling $_____

($5.50 per set in U.S. Overseas surface shipping add $10.00 per set. For Air Service to Alaska, Hawaii, Canada, Mexico and Central America, add $22.00 to book amount. For Air Service to all other foreign countries, add $32.00.)

TOTAL $_____

Name/Title _____

Station/Company _____

Billing Address _____

City/State/Zip _____

Phone/Fax _____

Shipping Address _____

Checks to: Streamline Publishing, Inc. Amount enclosed $_____.

Charge it to: ☐ MC   ☐ VISA   ☐ AMEX

Card # Exp. Date _____

Signature _____